The Mind-Body Problem and Its Solution

(Second Edition)

Carey R. Carlson

The Mind-Body Problem and Its Solution (Second Edition)
Copyright © 2019 by Carey R. Carlson

All rights reserved. No part of this publication may be reproduced, distributed, or transmitted in any form or by any means, including photocopying, recording, or other electronic or mechanical methods, without the prior written permission of the publisher or author, except in the case of brief quotations embodied in critical reviews and certain other noncommercial uses permitted by copyright law.

Although every precaution has been taken to verify the accuracy of the information contained herein, the author and publisher assume no responsibility for any errors or omissions. No liability is assumed for damages that may result from the use of information contained within.

Library of Congress Control Number: 2019917727
ISBN-13:   Paperback: 978-1-950073-92-4
            epub: 978-1-64749-003-4

Printed in the United States of America

GoToPublish LLC
1-888-337-1724
www.gotopublish.com
info@gotopublish.com

Contents

Preface vi

Synopsis by Chapter viii

Introduction—Minds and Bodies xi

References xv

Chapter 1 The Presence of Sensory Qualities 1

Chapter 2 The Absence of Qualities in Physics 16

Chapter 3 The Mind-Body Problem 28

Chapter 4 Relations and Structure 39

Chapter 5 Space-time as Causal Structure 48

Chapter 6 The Physical Location of Mental Events ... 76

Chapter 7 Scientific Knowledge Characterized 94

Chapter 8 The Solution 114

Acknowledgements 149

Preface

THE MIND-BODY PROBLEM demands a description of how the mental and physical parts of the world go together to make up the whole. The problem was solved around 1927 by Bertrand Russell and Alfred North Whitehead. The solution involves a change in our conception of the so-called "material world." Ancient animistic views ascribed lifelike, personal characteristics to the forces of nature. Science has promoted a contrary view, in which the world consists entirely of geometrical fields and particles, devoid of feeling. The solution to the mind-body problem reconciles the genuine content of scientific knowledge with the essential nature of mind—its subjective feeling and its wealth of sensory qualities. The new understanding reverts in large part to the pre-scientific intuition of nature. Every quantum event of physics is an instance of subjective feeling—an "occasion of experience." When one such occasion influences another, a causal relation is exemplified. The scientific method discovers patterns of succession due to this causal relation, and scientific knowledge pertains exclusively to such patterns.

An elaboration of the foregoing provides a coherent understanding of the relation of mind and body-- in particular, the relation of the human mind and brain. At the same time, the method of science is clarified, as well as the nature of the information obtained by that method. I will document the solution as provided by Whitehead and Russell. Their writings provide the several points of understanding needed to correct the prevalent misconception of the physical world. I will lift from their writings just those ideas that are crucial for making clear the relation between mental and physical entities.

The problem and its solution are laid bare in the opening synopsis. The rest of the book serves to make the meaning of the synopsis unmistakable. No special preparation in philosophy or science is required.

The reward of gaining a more coherent view of the world goes beyond the immediate puzzle-solving pleasure. Matters of religion and one's place in the universe are recast in the light of more adequate fundamental concepts.

Synopsis by Chapter

1. **The Presence of Sensory Qualities.** An essential aspect of mind is the presence of qualitative sensory characteristics, such as colors, which provide recognizable feeling and experience. The restricted notion of mind as *feeling* is called "sentience." The recognizable sense qualities, known by immediate acquaintance, allow us to describe the variety of our directly felt experience. Such description, known as "phenomenology," is independent of, and prerequisite for, knowledge acquired through the methods of physical science.

2. **The Absence of Qualities in Physics.** Science has refined our notion of bodies such that the human brain and body are sub-systems of a few fundamental forces that account for the entire universe. These forces are defined purely in terms of mathematical quantity and structure. Qualitative sensory characteristics are absent in the finished theory. Bodies, particles, and fields are extended in space and exist for specific periods of time, without phenomenological qualities and without the sentience that depends upon such qualities.

3. **The Mind-Body Problem.** Science culminates in a theory of particles and forces that excludes the qualities of sentient experience. That being the case, sentient qualities and sentient experience, which seem at the outset to be an integral part of nature, are instead relegated to a parallel existence beyond scientific explanation. This radical dissociation casts doubt on our basic concepts of "mental" and "physical," and this is the mind-body problem.

4. **Relations and Structure.** Relations account for whatever order and structure are to be found in any realm of investigation. Relations and structure are among the phenomena presented to our sentient minds. Relations and structure form the basis of mathematics, and together with causal assumptions, the basis of physics.

5. **Space-time as Causal Structure.** Special Relativity eliminates instantaneous spatial relations in favor of time-ordering causal relations. Causal relations are definable without recourse to geometric notions. Time order, for physics, is relative position in a causal chain of events. Two events not ordered by a causal chain are called "contemporaries." Spatial order is defined for contemporaries by the convergence of their respective causal chains at common causal ancestors and descendants.

6. **The Physical Location of Mental Events.** Mental events have physical location by the same criterion as physical events, strictly by the theory of their causes and effects. Mental events are between their causes and effects, and this causal positioning is the complete criterion and meaning of their physical location, as it is for events in general, mental or non-mental.

7. **Scientific Knowledge Characterized.** Physical science constructs a causal model of the world for better predicting the patterns of qualities witnessed in human mental experience. The scientist has no privileged capacity to escape the confines of his mind to investigate the physical world directly. A predictive model is framed, tested, and refined solely based on phenomena witnessed in mental events. Scientific knowledge resides entirely in such models.

8. **The Solution.** Science delivers only the bare causal pattern of events. Among these events are sentient occasions of human perception, which provide science with its observational data. When the remaining events required for the causal pattern are considered sentient occasions also, a coherent view of the world is obtained.

Introduction—Minds and Bodies

"Mind" and "body" are basic notions we have of things that exist for specific periods of time in the actual world. Together they account for the stuff of everyday reality.

When you die, your brainwaves stop. When your brainwaves stop, you are pronounced dead. Your body may be kept alive, but only in a vegetative state. It is apparent that, in some sense, your brainwave activity *is* your mental experience. It is the invariable accompaniment to your conscious existence. A shift to lower frequency signals a condition of deep sleep, while alpha frequencies characterize periods of waking and dreaming. This correspondence seems natural under the assumption that some component of brain activity, and human mental experience, are one and the same thing.

We hit a snag in this easygoing identity, however. It is contradictory to assert that two things are identical to one another if they differ intrinsically from one another. The sensory qualities that characterize our mental experience are, on the scientific account, no part of the physical world—a situation prevailing since the time of Newton. This prevents the identification of any recognizable feature of our mental experience with any feature of the physical world. Our sentient minds and our physical bodies are consigned to parallel worlds, related only by an unexplained coincidence in time. This dissociation between mind and body is repugnant to anyone who fully grasps it, which drives some to attribute the harmony of mind and body to the unknowable power of a deity. That strikes others as a premature surrender of rationality—that mind-body difficulties are more likely due to our own mistaken assumptions. Prominent assaults on the problem treat one or the other of "mind" and "body" as a mistaken or confused notion, but none of these has proved convincing. The average person believes in the reality of both mind and body and does not suppose that philosophy has made any great strides beyond common sense in explaining their relationship, though perhaps science has.

Nearly everyone though, has entertained the problem in some form. When a tree falls in the forest and there's no one there to hear it, is there a sound? Could a computer become conscious? Could personal awareness survive the death of the body? These inquiries come to a common impasse at the classical mind-body problem.

...

If a tree falls in the forest, and there's no one there to hear it, is there a sound? If "sound" is taken to mean a qualitative phenomenon characterizing mental experience, then no, there is no sound. If "sound" is taken to mean the mechanical vibrations of air, then yes, there is sound. So, by removing a major ambiguity of the word "sound," we can put the question to rest and go about our business.

We can settle a similar issue by distinguishing two meanings of the word "color." If sunlight filters through the leaves in the Amazon jungle, and there's no one there to see it, is there color? One finds that each of the five senses gives rise to a vocabulary for qualitative, recognizable features of reality. The words of these vocabularies invariably have alternative meanings that derive from the scientific view of the world. In this view, the systematic causes of our various sensory experiences are depicted, ultimately in terms of quantity and mathematical structure, as for instance, pressure waves, electromagnetic waves, and frequencies and pathways of nerve impulses. There is a consistency and completeness in the mature framework of modern science, expressed in terms of numbers, variables, and equations. In that description of the universe, none of the qualitative properties of human perceptual experience are ascribed to the forest. The scientific forest is one of quantitative energy transactions, to which science ascribes no qualities.

Apparently, we can make short work of sorting out the ambiguities of the whole class of words that arise from sensory experience, not just "color" and "sound." We can answer the whole question of what *is* present in the forest (energy fields), and what is *not* present in the forest (the felt qualities of sensory experience) when no one is there to witness the events. The questions lead to a straightforward resolution.

But suppose there *is* someone present to hear the falling tree. Then surely there is *sound* in every sense of the word. But the scientific description of the situation remains in principle unchanged. The human brain and body are no exception to the mathematically expressed theories of physics. They're made of the same stuff. No new fundamental fields of force are introduced. The observer's perceptions, verbal reports, and entire physical existence are accounted for by mathematical complexities in the energy fields. To account for the qualitative sound heard in the forest, we must recognize something in addition to the observer's brain and body. We must recognize the qualities themselves, as given in our sentient experience. In short, we must recognize the observer's mind. Only then do we have a forest that includes sound in every sense of the word.

...

A person has a mind, and a person has a body. How deep does the distinction go? Is it possible for your mind, your "stream of consciousness," to survive the death of your body and brain? Can we conceive of such a disembodied stream of consciousness existing in time, consisting of nothing more than its own thoughts and feelings? As for physical bodies, we all believe that they can exist independently of our minds. And we shall pursue the working scientific assumption that the human body, like any body, consists entirely of electromagnetic, nuclear, and gravitational energies. If the human mind can be conceived, in terms peculiar to it, sufficiently complete to count as something existing in time, and if the same holds true for a physical body, then we should say the distinction between mind and body goes very deep.

If, on the other hand, supplied with nothing but the fundamental notions of physics, we can conceive of the stream of consciousness solely in terms of electromagnetic processes in the brain, then the distinction between mind and body does *not* go very deep. The human mind is especially bound up, as a matter of empirical finding, with certain electromagnetic activities of the brain. If the notion of "the mind" could be broken down into the fundamental notions of physics and supplanted by them, then the mind would be eliminated as a kind of existence distinct from the body. The mind would be just some component of physical energy. In this case, mind and body present no incoherence, and there is no mind-body problem.

Our aim in the first three chapters is to examine the essential characteristics of minds and bodies, as ordinarily conceived, which have made the relationship between them a genuine mystery, the core problem of philosophy through the ages.

REFERENCES

All quotations and page number references are from HK, AM, or AI.

HK: *Human Knowledge, Its Scope and Limits*
by Bertrand Russell
Copyright 1948 by Bertrand Russell
A Clarion Book
Published by Simon and Schuster
Rockefeller Center, 630 Fifth Avenue, NY, NY 10020
Third paperback printing, 1967
Printed by Murray Printing Co., Forge Village, Mass.

AM: *The Analysis of Matter* by Bertrand Russell
Copyright 1954 by Dover Publications, Inc.
(republication of the original 1927 work)
Dover Publications, Inc.
180 Varick Street, New York, N.Y. 10014

AI: *Adventures of Ideas* by Alfred North Whitehead
Copyright 1933 by The Macmillan Company
Copyright renewed 1961 by Evelyn Whitehead
Collier-Macmillan Canada, Ltd., Toronto, Ontario
First Free Press Paperback Edition 1967

QED The Strange Theory of Light and Matter
by Richard P. Feynman
Copyright 1985 by Richard P. Feynman
Published by Princeton University Press
41 William Street, Princeton, New Jersey 08540

CHAPTER 1

The Presence of Sensory Qualities

An essential aspect of mind is the presence of qualitative sensory characteristics, such as colors, which provide recognizable feeling and experience. The restricted notion of mind as feeling is called "sentience." The recognizable sense qualities, known by immediate acquaintance, allow us to describe the variety of our directly felt experience. Such description, known as "phenomenology," is independent of, and prerequisite for, knowledge acquired through the methods of physical science.

An Unsettling Dream

You wake up to the sound of the alarm clock. You get up, get dressed, and eat breakfast. Leaving the house, you pick the newspaper off the doorstep and get into your car. You turn the ignition key… and suddenly you are in bed waking to the sound of the alarm. Sure enough, you're under the covers and the alarm is ringing. It's fresh in your memory that you just dreamed of waking up and heading off to work. You turn off the alarm and contemplate the oddity of the dream. You get dressed for work and eat some breakfast. On your way out, you pick the newspaper off the doorstep. You climb into your car. You turn the ignition key and the car starts. You recall that your dream had ended when you turned the ignition key. On the way to work, you're waiting at a stoplight. Just as it turns green... the alarm rings. You're back in bed, staring at the ringing alarm. You grab it and throw it against the wall. What if you're still dreaming? This latest wake-up seems real enough, but no more real than the last one. What if you're not yet awake? You're not quite sure.

You get out of bed, feeling shaky. What if you've gone insane? You light a cigarette and make some coffee. You phone a friend, confessing to an "anxiety attack." Your neighbor is reassuring-- he'll be right over. While waiting for him, you make some frantic observations around the house. You examine yourself in the mirror. You turn on the TV to check what's being broadcast against the TV guide. Everything checks out. By the time your friend arrives, you're embarrassed to have called him. A dream, or a series of dreams, has merely confused you. Your neighbor listens to your account of the dream. He tells you that such an experience would upset anyone. He offers to drive you to work. You hesitate but accept. By the time you get to the office, the old confidence is back. You sit down at your desk, and the phone rings. No, it's the alarm. You're back in bed, just waking up...

I heard that story in a philosophy class from Professor Keith Gunderson. The story illustrates something about dreams, something about waking life, and what these two have in common.

Regarding dreams, we may say that they can present us with such sights, sounds, tactile impressions, and apparent interactions with other people as to constitute a full-blown but illusory experience, which usually fades rapidly upon waking up.

Regarding waking life, it can be concluded that confidence in the trusty world of waking life rests entirely upon consistency checks. These checks generally secure a practical certainty before doubts even arise. However, to grasp the intent of the story is to understand that these consistency checks guarantee something less than logical certainty.

And what then do dreams and waking life have in common? They both involve a range of sense impressions and qualitative states of mind. Let us call such a state of mind, with its range of sense impressions, whether it falls into the context of a dream or into the context of waking life, a human "sensorium." You have a sensorium when you are dreaming, and you have a sensorium when you are awake.

We shall use the word "sensorium" to refer to a mind, insofar as a mind consists of sense, sentience, or feeling. Although one discriminates a qualitative variety of feelings within a given moment, "sensorium" conveys the fact that the elements of this variety form a unified whole. A sensorium, over time, is a "stream of consciousness." I'm merely labeling something that everyone has—something ever-present and taken for granted. Having adopted the peculiar word for the ordinary thing, we can, when we like, avoid more general-purpose terms that have taken on multiple uses and ambiguities.

Why not just use the word "mind" and be done with it? After all, it's the mind-body problem we are presenting here. For one thing, mind includes *unconscious* processes, the theory of which is both contentious and tangential to our core subject. We will proceed without unnecessary difficulties by restricting our focus to mind as sentient awareness. With this restriction, to exist mentally—to have a mind—is to be sentient, to *feel*.

Mind also involves *intelligence*. Along that line, one can consider "smart machines," and the "mentality" that might consequently be ascribed to computers. We will find the core of our problem in the consideration of simple "raw feeling," without exploring the difficulties of higher versus lower mentality.

Finally, mind is the domain of motivation or *purpose*, as opposed to the mechanistic causation of physical science. We don't need to grapple with that distinction either, in order to present the paradoxical coexistence of a sensorium and a brain. Once that difficulty is laid bare, we will proceed with its resolution. We will then have course to a natural explanation for the purposeful nature of mind, and the role of such purpose in the physics of cause-and-effect.

In this chapter we are just trying to whittle our attention down to a person's mental existence as a sensorium, characterized entirely in terms of feeling or sentient awareness. This restriction, let us note, does not reduce our field of study to scant nothing, since the sensorium includes the feeling of self-awareness, which is generally revered as an intractable mystery when the topic of discussion is "consciousness and the brain."

...

"Phenomenology" is the descriptive characterization of one's immediate experience, without venturing beyond what is directly presented. The attitude is taken that appearances are worthy of examination for what they are in themselves. They are not just indicators of a wider realm beyond the field of sentient awareness. Phenomenology describes what is openly disclosed in sentient experience, surveying a qualitative realm of features and patterns. Phenomenology is the study of the sensorium.

Think about what redness is, in itself—that is, without regard to the theoretical entities of physics involved in its causation, such as electromagnetic energy, its absorption and reflection at various surfaces, and subsequent excitations in the eye and brain. The quality red has no description in terms of mass, charge, frequency, or motion. You're left with redness itself to contemplate. Focus your attention upon redness as an essence. It is both irreducibly simple and strongly identifiable. This essential nature of redness is apparent to your mind when you focus your attention upon redness as the unmediated sensory quality that it is. This is redness as a "phenomenon." When you adopt the phenomenological attitude toward the directly given features of experience, there occurs a temporary suspension of interest in the wider physical realm believed to underlie the surface of appearances. This wider realm, which is the usual focus of daily living, is said to be "bracketed out" of consideration when you adopt the phenomenological attitude.

As a simple, recognizable characteristic, presented directly and repeatedly to awareness, redness can be termed an "object" of awareness in a general sense. Let us call the act of awareness, within which redness makes its appearance, a "sentient experience." The act of awareness, and its objects, require one another. To exist as a sentient being is to have specific qualitative feelings. Without some characterizing phenomenon, there is no experience, and sentient mind does not exist. Sensory qualities are the very medium of mental experience.

When any feeling occurs there is sentient experience, and the phenomena that present themselves in that experience form a whole. A moment of sentient experience can be considered a basic type of temporal existence for the purpose of describing the actual world. Sentient experience, and the phenomena presented within it, are investigated in the discipline of phenomenology. This discipline uses an introspective method that is not dependent on the method, or the results, of modern science.

...

Let's take a time-out. What are we trying to accomplish here? There may be some readers for whom this is too obvious, and others for whom no meaning is accruing whatsoever. It may help to explain that in philosophy, where common sense categories of existence are subjected to uncommon standards of logical rigor, the realm of the mental, which I am trying to define in this chapter for the purpose of subsequent discussion, is questioned or denied by many philosophers. Common sense notions about mind are re-construed in such a way as to allow for the practical usefulness of mental terminology while restricting the ultimate description of reality to the terminology of physics. This involves the denial of sensory qualities, sentience itself, and anything purely mental. Contemporary physics is promoted to a complete theory of all that there is. The mind-body difficulty is solved by the removal or replacement of all things mental.

In 1970 I'd had some college-level background in science, but no courses in philosophy. Someone gave me a copy of Whitehead's *Science and the Modern World*, which showed me that redness is something in itself. Here is Whitehead again, in *Adventures of Ideas*.

> Gaze at a patch of red. In itself as an object, and apart from other factors of concern, this patch of red, as the mere object of that present act of perception, is silent as to the past or the future. How it originates, how it will vanish, whether indeed there was a past, and whether there will be a future, are not disclosed by its own nature. No material for the interpretation of sensa is provided by the sensa themselves, as they stand starkly, barely, present and immediate. We do interpret them, but no thanks for the feat is due to them. (AI, 180)

Yet redness is not included in the concepts of physics. I had somehow acquired, without being aware of it, an outlook dominated by physics, which overlooked the realm of sensory qualities. These qualities had taken on a subliminal role in my awareness, as mere cues to more substantial objects. I was suddenly struck by the immaterial nature of the color red as though by a paranormal experience. This disruption of my complacent view of the world might never have occurred if I had not happened upon *Science and the Modern World*. Philosophers who promote a view that denies the existence of mental phenomena can have more success than one might suppose. Modern culture is imbued with respect for the superiority of the scientific view of the world. A person today can acquire a physicalistic view of the world without even knowing it.

Someone involved in the visual arts, whose stock-in-trade is visual form and color and their impact on the human mind, is apt to find the present chapter too obvious for words. The same reader may puzzle at the physicalist view of the world presented in Chapter 2, although that will be equally rudimentary. Each of the first two chapters covers something simple, fundamental, and extremely general. And therein lies the difficulty.

How many people have you met who have wrestled with the mind-body problem? None, perhaps? Literate society in the time of Newton and Descartes were aware of the problem, but it went unsolved for too many centuries. To shield ourselves from a humiliating lack of progress, we now look over, under, around, and through the mind-body distinction, without looking *at* it. Our view of the world is bifurcated, like the vision of a fish whose two eyes see nothing in common. We must force our isolated intuitions of mental and physical, however incongruous, into a mutual encounter, as Descartes did so well several centuries ago. We will then be in position to follow Russell and Whitehead to the solution.

One might expect academic philosophy to herald the solution as its greatest achievement and a new beginning, but that has not happened. Academic philosophy today is not generally aware of any solution. Attempts to "explain away" mental entities still garner much interest. But we will take the existence of mental entities as a simple fact and shall find no need to sweep them under the carpet. This chapter is nothing but an ostensive definition of sensory qualities—that is, an assignment of terms like "redness" to denote identifiable qualities involved in sensory experience. This may not seem like much of a basis for establishing the existence of mind, but it will suffice, together with the next chapter, to present the mind-body problem.

We continue then with our aim of drawing attention to the realm of phenomena, appealing to the deliverances of the reader's own senses. We adduce the body of literature classified as "phenomenology" as having that same endeavor, with a wealth of descriptive results to show for it. We will go beyond examples of utterly simple phenomena, such as redness, to facts of modest complexity—facts situated entirely within the province of phenomenology. To the mere naming of qualities we shall add statements of fact, which are only verifiable within the phenomenal field of awareness. The purpose is merely to fortify the belief that sensory qualities and the patterns they exhibit, ephemeral though they may seem in comparison to physical bodies, are not mere nothing.

...

Let us now consider, as an example of a purely phenomenological fact, one specific way in which colors are related to one another—namely, that a sufficiently complete set of hues forms a "hue circle." I want to make it clear that this fact is itself a phenomenon, established entirely through phenomenological means.

When different colors are seen simultaneously, their similarity or dissimilarity can be judged, and furthermore, the degree of similarity can be judged. If many colors are presented in many arrangements in the visual field, certain smooth arrangements exist such that any two neighboring colors are very similar. A hue circle of colored patches is one such arrangement. The progression of hues around the circle is in rainbow order, except that red and violet, at the two ends of the rainbow spectrum, are joined by magenta hues missing from the rainbow to complete a circle. Among individual colors in the hue circle, primaries can be chosen such that any two of the primaries are as dissimilar from each other as color hues can be, and any other hue can be described by its degree of similarity to the neighboring primaries on either side of it in the hue circle. Red, yellow, green, and blue form such a set of primaries, the listing obeying their order in the hue circle. There are reddish yellows, yellowish greens, greenish blues, and blue-ish reds, but there is no such thing as a reddish green, or a yellowish blue. These are facts inherent in the hue circle itself, having entirely to do with the intrinsic nature of color.

It is also a phenomenological fact that the totality of colors, when a geometric analogy is employed, forms an irregular 3-dimensional shape known as "the color solid." When "just noticeable differences" are used to arrange colors geometrically, an oblong 3-dimensional form results. A dark-to-light series of gray tones running through this color solid serves as a linear axis which supplies the gray at the "hub" of any cross-section containing a hue circle. Hue, saturation, and lightness are three inherent dimensions of color, coordinated by the scale of gray tones as one reference axis. When black and white, at the ends of the gray scale, are included as phenomenological primaries along with red, green, yellow, and blue, any given color can be described by its degree of similarity to each of the six primaries.

The foregoing facts are innate to the realm of color. They are not trivial to formulate. A certain amount of mathematical intuition and study of colors is needed. But science as we know it has nothing to contribute to the study. Rather, the facts are founded without scientific assumptions. They are "givens" to which scientific theory must conform, if science should decide to take an interest. These facts could be established and understood by a perfectly pre-scientific culture. It is knowledge ascertainable only by direct acquaintance with colors.

> When we consider sound qualities, we do not find a ring-like structure that is a direct counterpart to the hue circle of the color realm. The dimension of pitch, which orders tones by higher and lower, forms a linear series. There are octaves which form natural cycles in the series of pitch, and other recognizable intervals that relate pitches. Loudness and tonality add further dimensions. Recall a favorite tune. In its musical essence, it is a

phenomenological pattern of various pitches of sensed duration, loudness, tonality and tempo.

Taste and smell sensations admit to some ordering on an intensity scale, but otherwise their qualitative variations are not susceptible to analysis into dimensions. If they exhibit a messy variety of qualities not amenable to categorization, they are nevertheless something, and help to sustain sentience in the event of deprivation of other senses.

Sensations of touch include a site of location within or upon a person's body image. This body-as-felt—the *phenomenological body*—is a structure of touch and kinesthetic sensations, and includes, for example, the limb that the physical body lacks in a case of "phantom limb" phenomenon.

When taken in combination, the various senses produce further complexities of variety and pattern in the sensorium. In fact, they provide a sufficiently rich totality to be commonly mistaken for the physical environment beyond the human head. (That this is a mistake is taken up at the outset of the next chapter.) In the case of waking life, the mistake does no apparent harm, a useful correspondence obtaining between the display of sensory qualities and the state of physical surroundings. In the case of dreams, the motor system of the body is disengaged, which prevents physical acting out. The confusions of hallucination, on the other hand, are dangerous, because a person's bodily actions are apt to be severely misdirected by a play of appearances in the sensorium that bears little correspondence to the physical environment.

...

Consider the notion of "virtual reality" as depicted in many science fiction movies. In the film *The Matrix*, the hero, along with most of humanity, is deluded in the belief that his sensory experiences constitute veridical perceptions of the world outside his head when in fact they do not. An alien race (of "sentient programs") is fiendishly manipulating nerve impulses in the hero's brain. The contraption which wires his brain into a computer-generated virtual reality is replicated on a mass scale to do the same for the whole human race. Unknown to this hapless lot, their physical bodies lie *in vitro* in life-support pods. This setup allows a coordinated deception of the entire human society, so that each person's virtual reality is coordinated with others, including a convincing "feed" of sensations that suggest a familiar physical environment. Today's theater-going public has no difficulty in following the story line that moves back and forth between the virtual world and the real world, thanks to proven techniques of cinema. The same cuts and techniques work well to convey, from a first-person point of view, a character experiencing memories, dreams, or hallucinations. No magic is required from the celluloid beyond its basic trick of reproducing sounds and images.

The science fiction premise of The Matrix is neatly in accord with brain science, differing only in respect to state-of-the-art limitations. Normal veridical perception also involves a qualitative *virtual reality* occurring in our heads, but presumably without any villainous piracy of our nerve signals. A wide external reality is judiciously represented in each of our waking virtual realities, fulfilling our beliefs that we act and communicate meaningfully in a common social world. We understand the premise of The Matrix without difficulty, by intuitive appeal to the sensorium as the common element in dreaming, hallucination, and waking life. This sensorium exists whenever and wherever sentient experience of any form occurs. The sensorium is implicit in our understanding of virtual reality.

To synthesize virtual reality, helmets and wires are used to intervene in the communication between a person's sensorium and the rest of the world. That involves the scientific account of perception, covered in the next chapter.

...

Now let's expand the sensorium beyond the five senses. What of the feeling of mental anxiety? What of the feeling that you know something? What of the feeling that you doubt something? What of the feeling of harboring intentions, and engaging in their fulfillment? These are recognizable feelings that are perhaps not reducible to the qualities of the five senses, even when taken in combination. We shall include them in the sensorium as further phenomenological constituents of sentient experience.

Whitehead recognizes feeling that supplements the five senses, which he calls "non-sensuous perception." As an example, he describes a very short-term type of memory which is an intimate feature of experience.

> Non-Sensuous Perception.— ... In human experience, the most compelling example of non-sensuous perception is our knowledge of our own immediate past. I am not referring to our memories of a day past, or of an hour past, or of a minute past. Such memories are blurred and confused by the intervening occasions of our personal existence. But our immediate past is constituted by that occasion, or by that group of fused occasions, which enters into experience devoid of any perceptible medium intervening between it and the present immediate fact. Roughly speaking, it is that portion of our past lying between a tenth of a second and half a second ago. It is gone, and yet it is here. It is our indubitable self, the foundation of our present existence. Yet the present occasion while claiming self-identity, while sharing the very nature of the bygone occasion in all its living activities, nevertheless is engaged in modifying it, in adjusting it to *other* influences, in completing it with *other* values, in deflecting it to *other* purposes. The present moment is constituted by the influx of *the other* into that self-identity which is the continued life of the immediate past within the immediacy of the present. (AI, 181)

A person's feeling of "now" is designated at the end of the quote as "the immediacy of the present." It is difficult or impossible to pin down a precise "present moment" of experience, but Whitehead succeeds admirably in describing the experience of temporal flux pervading the present moment. This felt passage of time cuts across the divisions of the various senses, uniting them into a common temporal stream. The ordering of one's experiences into a serial "stream of consciousness" is implicit in the sensed relation of before-and-after.

Equipped with the sense of time, we can classify various activities of mind as phenomenologically given. Mental life includes thinking, remembering, deciding, intending and all such mental acts and activities. The feelings involved are less vivid perhaps, than color qualities presented in the visual field, but we would impoverish our account of the sensorium if we fastened only upon qualities that differentiate the five senses. Indeed, as Russell says, "What we directly experience might be all that exists, if we did not have reason to believe that our sensations have external causes. ..." (HK 311). Russell denotes by "what we directly experience" that which phenomenology is intent on describing, and what we are here calling "the sensorium." Phenomenology, as a discipline, shades off smoothly into human experience in general, which is taken up entirely with the direct witnessing of phenomena, but not generally with the aim of disciplined, systematic description. I have stated, somewhat defensively, that phenomena are not "mere nothing." A more aggressive proponent might claim that phenomenological experience is all that exists. Physics, as we shall consider in the next chapter, tends to the competing claim that its own mathematically characterized world is all that exists. The mind-body problem could be called "the phenomenology-physics problem." The solution to the problem must adjudicate these competing claims over all that exists.

Neither Russell nor Whitehead relies on the prominent phenomenologists of their time. Russell and Whitehead do their own phenomenology, borrowing from older traditions. Some phenomenologists have analyzed a sentient experience as a *mental act* which *intends* its objects. Whitehead's *occasions of experience*, and the objects *given* to them, are at least roughly equivalent to mental acts and their intended objects. Russell sometimes speaks of the "I-Now" of experience as an irreducible entity which binds various phenomena into the unity of a mental event. Again, this seems roughly consistent with mental acts and their objects. Regardless of disputes within phenomenology over how, or whether, the unity of a mental occasion is to be analyzed into subject and object, I submit that phenomenology, as the descriptive analysis of sentient mind, exists alongside physical science as a peer in the basic categories of human knowledge.

There was a philosophical paper titled "What is it Like to Be a Bat?" The title alone prompts a person to imagine a bat's experience, borrowing initially from the sense of what it is like to be a person, but anticipating a major adjustment for the bat's sonar sense. In any case, most people would assume that it is like *something* to be a bat, or, in other words, that bats have feelings. Most people assume that their pets have feelings—that they sometimes feel pain for instance. Pain is a phenomenological quality *par excellence*, and as such, a component of sentient mind. Common sense supposes that pain is felt by the higher mammals at least. In the present context, the ascription of sentient mind to these higher mammals implies nothing beyond what common sense already believes. We are just making further use of "sentient mind" to make its meaning clear. In that spirit, our pets are sentient, and phenomenology, although a strictly human activity, discloses the sort of thing that human beings routinely generalize to animals as well.

When pressed to draw the line, common sense supposes that sentience comes into being with the biological evolution of life at some minimal stage of complexity. Before that, apart from a possible Creator, the world is thought to be the interplay of blind forces devoid of feeling. That these physical forces, by arriving at a suitable configuration, can conjure sentience into being, is a magical tenet in the blind spot of common sense. Setting aside the question of how sentient qualities could be produced by physical forces, let us agree with common sense that nature somehow endows human beings and various animals with phenomenological feeling.

...

This concludes my attempt to define "sentience," or "sentient mind," and the qualitative phenomena that comprise it. If you can readily make conceptual reference to some feature of your own sentient experience, such as a favorite color, a familiar tune, or the feeling of the present moment, the purpose of this chapter is fulfilled. You can frame for consideration a definite phenomenon revealed to sentient awareness, uncontaminated by any scientific doctrines regarding the physical composition of the world.

CHAPTER 2

The Absence of Qualities in Physics

> *Science has refined our notion of bodies such that the human body and brain are subsystems of a few fundamental forces that account for the entire universe. These forces are defined purely in terms of mathematical quantity and structure. Qualitative sensory characteristics are absent in the finished theory. Bodies, particles and fields are extended in space, and exist for specific periods of time, without phenomenological qualities, and without the sentience that depends upon such qualities.*

Science has established that an observer has no direct perception of the world outside his own head. The observer can no more extend his direct perception beyond this range, than he can return backward in time. Events outside the head are *indirectly* perceived from within the brain due to mediating events such as nerve impulses, which elapse in a sequence of measurable time intervals. When you detect an event outside your head, a series of events has transmitted some effect to you. The duration of the mediating events imposes a separation in time between the external event detected and the subsequent detecting event that occurs in the head. The meaning of the term "indirect" as applied here to perception refers to cases in which causal chains of distinct events intervene, in the scientific account, between the initiating external event and the subsequent "percept," or perceiving event, inside the head.

The fact is that you cannot "get outside your own head" to perceive directly what anything outside your head is like, given the scientific basics of human perception. But this requires you to abandon a gut-level belief to the contrary. We all start out interpreting the deliverances of our senses as a direct revelation of what the world beyond our bodies is like. We assume that we have a direct and open portal on the world. In this we are instinctively and confidently wrong. To take a specific example, the scientific conception of the world attributes no color at all to the world outside the human head. Electromagnetic entities and events, defined purely in terms of quantity and spatial-temporal location, connect with more entities and events of the same kind within the human head. Only then, when terms describing human mental experiences are abruptly introduced, does real color enter the description. We persist in thinking that colors are properties of physical objects, and located in the space extended beyond us, where we reckon the physical objects to be. But science has no use for the idea that the color red travels from an apple, through space, to our eyes. The description of the traveling influence in this case is already complete in terms of the quantitative theory of electromagnetic radiation, which does not employ qualitative color, and finds no room for it in its explanation.

Physics has refined its laws with ever better predictive results, to the point where the defined entities obeying these laws are never directly perceived in the sentient experience of a human observer. In that regard, all the entities dealt with in modern physics, including tables, chairs and human bodies, are strictly theoretical constructions, defined entirely by systematic conjecture. Physical bodies are not phenomena. The coordinated visual images and touch sensations that make up our experience of tables and chairs should not be mistaken for the physical objects themselves. Physical bodies consist of quarks and the like, supposed by science to be without any sensual features that would allow them into phenomenal experience. We need to ask what legitimately remains, if anything, of the presumption that we know what physical bodies are like.

> Historically, physicists started from naïve realism, that is to say, from the belief that external objects are exactly as they seem. On the basis of this assumption, they developed a theory which made matter something quite unlike what we perceive. Thus their conclusion contradicted their premise, though no one except a few philosophers noticed this. (HK, 197)

Let's trace the development of the scientific notion of matter, in a fanciful account, to see how the earlier notion is contradicted at a later stage. We picture Galileo dropping two objects, a heavy one and a light one, off a tower, to settle a bet. The odds are favoring the heavy object to fall faster. But the two objects fall side by side, and land at the same time, the heavy object merely raising more dust. The experiment is repeated, and the time taken for objects to fall from various heights is measured more carefully. Eventually a formula is distilled which gives a single fixed rate of acceleration for any falling object, regardless of its weight.

At this point the customary notion of matter, and the scientific one, have not diverged. Matter is directly perceivable in its color, its shape, its trajectories, its resistance to being lifted or pushed around, and its hardness. Clocks, rulers, and scales, devised of this same stuff, are employed in discovering general laws of motion in direct observation experiments. Although the coloration of bodies has nothing to do with their motions, it makes them visible, which helps in observing the experiments.

The visibility of matter also helps Kepler, with the aid of a telescope, to observe the elliptical orbits of planets about the sun. He finds that a planet's arc sweeps out equal areas of an ellipse in equal time intervals. Newton finds a remarkably general hypothesis to explain this. Using Descartes' "Cartesian coordinates" to express geometry with algebraic formulas, and his own invented calculus, Newton accounts for falling bodies, the collision behavior of bodies, and the orbits of planets, all in a few simple formulas with units of mass, space, and time. Color is not among the primary variables and has no share in the explanatory power of science at this stage. It has an implicit role in the observation and verification of the theory, but color is not integrated into the new framework of science. It receives no benefit of explanation.

Electric and magnetic effects proved *not* to be reducible to Newton's laws of mass, space, and time. A new primary variable, "charge," must be added to Newton's three. Maxwell's equations of electromagnetism give the laws governing this new variable. It then becomes apparent that light, the presumed carrier of color from material bodies to the eye, is nothing else than a narrow band of frequencies in the spectrum of electromagnetic waves. The behavior of radiation in this band, first at the surface of a material body, then in transit through space, and finally at the eye, specifies the role of light in visual perception, using space, time, mass, and charge combined into formulas. Qualitative color is still not among the variables. The points of space, the moments of time, the quantities of mass and charge, are all without color. Color is no part of them, and they are no part of color. Colors are no longer considered to be within, or at the surface of, a material body. Nor is color any part of the propagating wave energy. Colors only arise as effects in the brain of an observer subsequent to the bombardment of the retinae by electromagnetic radiation. The retinae and brain, no exceptions to physics, also consist of colorless matter. Color, in the scientific picture, has lost its mooring in the physical world altogether. Color has no definite location in physical space. It is only in the mind.

To reiterate, material bodies are scientifically understood as having no color, and are thereby distinguished from the colored forms that appear in human visual experience. Colored patches are not located at the surfaces of material bodies. On the contrary, they are delayed effects in a causal sequence, "downstream" from the radiation events which emanate from distant colorless surfaces. But then, physics makes no new theory for what goes on inside the head versus what goes on outside. Colorless physical causes in the brain produce more colorless physical effects. It is left for the mind to host a private "colorized screening" of the world, while the eyes and brain engage in the colorless energy transactions specified by the laws of physics.

> Everything that we believe ourselves to know about the physical world depends entirely upon the assumption that there are causal laws. Sensations, and what we optimistically call "perceptions," are events in us. We do not actually see physical objects, any more than we hear electromagnetic waves when we listen to the wireless. ... (HK 311)

A change is consolidated in the scientific picture of nature by the time Newton and Maxwell have formulated their laws. "Nature" acquires its scientific definition in terms of its "primary properties," while the sensory qualities involved in mental experience are termed "secondary." And what are the primary properties, which now have the honor of defining the physical world? The primary properties are purely quantitative values assigned to space and time coordinates. The world is the full specification of these geometrically ordered quantities. The laws specify how one spatial configuration follows another in time. This is the stark scientific account.

...

How are we to frame an accurate concept of the universe in terms of its primary characteristics without recourse to the sensory qualities, which have already proven to be so misleading? For this, we need to distinguish "physical space" from the intrinsic geometric aspects of our visual, auditory, and kineasthetic experiences. After all, having acquired basic motor skills, it is the presupposition of everyday life that physical space exists all around us, regardless of the intermittent activation of our sensory fields. Therefore, physical space need not be considered a phenomenological entity at all. It may be considered an objective expanse of a geometric nature, shorn of sensory qualities and privileged perspectives. It furnishes the positions and directions of physical quantities. It bears only the mathematical character required for the formulation of the laws of physics.

If physics should require that spatial dimensions be added to the three that we have intuitively mastered, or that space be curved to better formulate the physical laws, our imaginations may be challenged, but we do not thereby despair of understanding science. Our familiarity with physical space, at the scale of our everyday activities, remains our basis for understanding what physical science is about. Science is about space and what's in it. We're not unduly alarmed to find that the physicists, at the scale of the very large and the very small, are tampering with the technical details of space and time. When confronted with something like "string theory," we take it to be some convoluted geometric conception, requiring a frightful mathematical imagination, but in principle, amounting to an elaboration of the three-dimensional space that is second nature to us all. Our understanding of science rests upon the intuition of physical space, and it seems impossible to doubt that we know what physical space is.

While it is widely held that modern physics has "dissolved matter," it is not so widely thought to have "dismantled space." We shall defer that line of inquiry in order not to disturb, at this point, the reader's basic intuition of "the physical world." This is the one artifice, from my point of view, needed to stage the mind-body problem as a problem, before explaining the solution. This is not for aesthetic effect. The first mental act, when one sets out to think about the natural world, is to call up an image of physical space, perhaps with something in it. Until that germinal intuition is shown to lead to paradox, a person simply will not call into doubt the basic premise from which all subsequent understanding of the natural world takes its departure. The solution to the mind-body problem will remain psychologically out-of-reach. That is why I suggest coming to grips with the problem before trying to understand the solution.

...

We are considering the notion of "physical," about which science claims the authoritative expertise. When Newton condensed the workings of the universe into a few deterministic laws of mechanics, the metaphysical view called "materialism" gained stature. This view held that Newton's conception of the material world provided a complete general description of all that exists. The mainstream of philosophy never accepted this, believing that some sort of mental or spiritual existence was left unaccounted for in the theory of matter and its motions. It can be said, however, that materialism served as the virtual blueprint for scientific progress for centuries to come. Materialism had to be drastically revised, not to accommodate the defenders of mind, but to deal with objections from science itself. Materialism, as the blueprint for scientific progress, led to its own drastic qualification. If it were to survive in a form compatible with science, it would have to do so without the notion of a substance called "matter."

Let us consider the "dematerialization of matter" in modern physics. This might be thought to solve the problem of mind-and-matter by disposing of matter. The notion of an inert particle of matter, which is without feeling of any sort, yet exists by virtue of its mass and location in space, is not a difficult one for most people. Rather, it is so un-difficult that it is nearly impossible to abandon. Nevertheless, particulate matter *has* been abandoned by science, since it does not bear scrutiny at the quantum level.

Tables and chairs still exist of course, but they do not consist of particulate matter, having given way to the probability waves of quantum mechanics. The notion of a particle of matter requires that it have a stable and instantaneous location in space, which was found to be incompatible with quantum theory. The latter theory has proved to be indispensable to physics, and so the notion of particulate matter has been discarded.

Since the inert particle of matter has been re-imagined by physics as a somewhat livelier packet of energy, it becomes possible to speculate that physical energy is somehow mental activity. But that remains an idle fancy so long as physical energy is simply conceived as that which is contained in physical space, while physical space is simply conceived as that which contains physical energy. With this pair of interlocked concepts at the core of our understanding, the physical world seems to be a geometric expanse of distributed stuff that requires no infusion of mental features to fortify its existence.

So long as we're satisfied that *something* in the theory of physics, such as a "force field," occupies the space where tables and chairs are believed to exist, the abolition of matter as such does little to change our beliefs about the physical world. We learn in the early grades that a material body is mostly empty space, the atoms being something like miniature planetary systems. That prepares us for the higher grades, where we find that occupied space contains not miniature planets, but immaterial quantum events. As the facts unfold, we rely increasingly upon the notion of physical space itself, while the notion of what physical space contains becomes less intuitive.

Today the "stuff" of physical science is quarks, gluons, leptons and so on. Quarks, for example, are posited in several paired types in order to build up a systematic model of the world from proposed elemental entities. The "building up" is accomplished in the theory by various transformations, combinations and calculations. These mathematical operations represent the co-dependencies and interactions of the elemental entities. The entities at the base of the construction are simply posited by conjecture—they are given arbitrary names to establish by fiat their bare logical distinction as individuals. In the same manner, they are further classified into posited types according to how they are supposed to form, in relation to other individuals, combinations of greater complexity. The constructing process must eventually arrive at assemblies that represent observable entities, so that the theory is capable of empirical confirmation. When a new construction is proposed, it must confront any rival theories. The new theory generally predicts, somewhere in its details, novel observations not predictable by rival theories. If the novel predictions are born out, the new theory gains acceptance. To win acceptance, a new theory must not only encompass the already established findings of science, it must also break new ground. In this manner, Relativity and Quantum Theory have replaced Newton's theory of mechanics, dispensing with Newton's ideas of time, space, matter, and fully deterministic laws.

How has the notion of "the physical world" changed as a result of modern physics? In important ways, the contemporary outlook remains the same as in Newton's time. As was the case with Newtonian theory, modern science confines its characterization of the physical world to mathematical descriptions. In hindsight, the essential role of conjecture in any theory of the physical world is now commonly acknowledged, since belief in "matter" could only have been wrong if it had been a conjecture in the first place. However, the belief that matter was devoid of feeling, or sentience, has carried over intact to the current conception of energy. Energy is now the insentient stuff that displaces matter as the occupant of space and time. This rough assessment is in line with the central thesis of this chapter, that sensory qualities have been thoroughly eliminated in the formulation of scientific theory. Since a notion of sentient experience cannot be framed without appeal to phenomenal qualities, contemporary physical theory is no more hospitable to the notion of sentient mind than was Newton's theory of matter. Descartes' analysis of the rift between mental experience and the physical world applies equally well to the dominant intuitions of today.

Are there not other sciences besides the physical sciences? Why not let the science of psychology deal with the phenomenon of human sentience, if physics is un-equipped for it? That suggestion meets a difficulty due to the unification of the sciences that has occurred as the specialized sciences have advanced. Science has effectively become a single monolithic theory, with physics providing all the fundamentals. Physics is aggressive. It does not curb the domain of its findings to accommodate the continuation of independent sciences. Chemistry is now understood to be founded upon physics, and in principle, is completely reducible to it. Biology is founded upon chemistry (and thus, upon physics) so that the problem of the origin of life, for example, is framed in terms of the proper conditions obtaining on the planet for an incubating "chemical soup." The physiology of the human brain and body is likewise subsumed entirely in the theory of physics. Psychology winds up as the repository for any difficulties associated with the mind-body problem. In the history of its development, psychology included an overt strain of phenomenology, which I have presented as an irreducibly mental realm. However, this whole phenomenological strain is suspect by today's standards of "hard science." The suspicion falls on the subjective reporting of privately observed contents of the mind. The qualitative characteristics described in such reports do not admit of the objective verification that distinguishes the physical sciences. If one ignores the intended references to mental phenomena reported by a human subject, in favor of the verbal utterances themselves, the role of the subject is confined to physically defined actions, such as lip movements when the subject speaks. This allows the human mind to be treated as a system of physical mechanisms, casting psychology as "the science of the brain." In this view, stimulus-and-response form an unbroken chain of physicalistic actions leading into, round and about, and out of, the brain. Psychology then secures the status of hard science founded upon physics, avoiding the contamination of phenomenological elements such as colors and pains.

In practice, psychology is still eclectic, and not conducted uniformly under the prescriptions of a physicalistic approach. Similarly, in the practice of medicine, a doctor sympathizes with pain for reasons that cannot stem from any training in physics. Any science-related effort that is focused directly on the well-being of human individuals is bound to involve a makeshift mind-body dualism. Physicalism has not yet succeeded in exterminating all belief in mental entities. I am framing the mind-body problem as a rift between the domains of phenomenology on the one hand and physics on the other. I suggest that psychology as a science, insofar as it straddles both domains, can at best patch together an amalgam of physical and phenomenological components. It must suffer, in its conceptual foundations, from the very same incoherence that we shall deal with directly, without presupposing any specific psychological theory.

Turning then to the ultimate basis of the physical sciences, how unified is physics itself? If Einstein had succeeded in formulating his "Unified Field Theory" we should answer that physics is completely unified. For some reason physicists changed the name of this theoretical goal to "Grand Unified Theory," and then again to "Theory of Everything." The efforts toward that goal, whatever its name, have shown inexorable progress. Physics was reduced at some point to gravity, electro-magnetism, and weak and strong nuclear forces. Since then, these four have been consolidated to two. That means there is one "seam" remaining in the fabric of physics, staving off the completion of theoretical science. At present therefore, unification is not complete, which indicates that something is wrong. But it seems to be only a matter of time until the right mathematical twist is found to express physics as a seamless theory of the physical world, complete in its own terms. (I will offer a theory in Chapter 5 as a final reduction of physics to time as the sole remaining parameter.)

Due to the unification of the sciences under physics, it is evident that mainstream science today frames a conception of the physical world entirely in terms of geometrically ordered quantities. The sensory qualities considered in the previous chapter are excluded at the outset from having any part in the theory. The same exclusion of sensory qualities was a feature of Newton's theory of matter in motion. While that theory has been drastically revised, the notion of the physical world as an insentient mechanism has been retained. This conventional summary shows that the contemporary scientific conception of the physical world contains no indication of any such thing as qualitative human experience, and though we persist in testifying to such experience, it is irrelevant, in the scientific view, to the course of physical events.

CHAPTER 3

The Mind-Body Problem

Science culminates in a theory of particles and forces that excludes the qualities of sentient experience. That being the case, sentient qualities and sentient experience, which seem at the outset to be an integral part of nature, are instead relegated to a parallel existence beyond scientific explanation. This radical dissociation casts doubt on our basic concepts of "mental" and "physical," and this is the mind-body problem.

The relevance of philosophy stands or falls with its ability to resolve the mind-problem, and success is not thought to be near at hand. In the meantime, society has lost interest. *Harry Potter and the Philosopher's Stone* was changed, for the American audience, to *Harry Potter and the Sorcerer's Stone*. The word "philosopher" was deemed the "kiss of death" for book sales and movie box office, indicating the esteem in which philosophy is held today. Yet the mind-body problem has points to commend it as an exciting read. It's an epic story, in which the towering figures of Plato and Aristotle, champions of mind and matter, contend with each other through the ages of western thought to forge an intelligible view of the world. The real-life mystery is well documented and dramatic. In the ending that I find compelling, Plato and Aristotle return to the stage as Alfred North Whitehead and Bertrand Russell to re-interpret the findings of modern science and remedy our distorted view of nature. Although there are elements of suspense and high stakes, the narrative concerns purely cognitive matters. These matters require a dispassionate analysis of our most fundamental concepts, and upon these, some logical hatchet-work to re-assemble a credible view of the world.

We have considered the fact that sensory qualities, such as colors, are present in experience. Indeed, there is no experience without such qualities. We then considered the fact that sensory qualities have been excluded from the framework of scientific theory. The joint implication of these two facts is problematic for the usual notion of "the natural world." We start out thinking that our sensory experience is mainly composed of sensory qualities, which is correct, and furthermore that these sensory qualities are part of the physical world, which now seems to be incorrect. If we take science seriously, and the increased use of mathematics that has been crucial to its development, we glimpse the world of science in its bare physicality, drained of all qualitative character. Our sensory experience, which owes its definition to sensory qualities, is the leftover residue. The physical world is truncated of sentient experience. Sensory experience is strictly in excess to scientific theory. It belongs only to the mental world described by phenomenology. The problem then confronts us as to how the mental world and the physical world can join to form, as we feel they should, one world.

By relying upon the sensory qualities to delineate the mental from the physical, we are traversing a well-worn path in philosophy. The problem in relating sensory qualities to physical entities is already in evidence in the contrast between Plato's *forms* and Aristotle's *matter*. The phrase "Platonic heaven" indicates the disconnect between the world of forms apprehended by Plato and the earthbound matter conceived by Aristotle.

Skipping to the dawn of modern science, we come to the point where phenomenology and physics, in their modern form, take independent paths. Descartes was so important to both phenomenology and physics that he could rightly be considered the founding father of each. On the side of physics, he fastened upon extension in space as the defining principle of physical existence, and he contributed the system of Cartesian coordinates that paved the way for Newton to express the laws of motion in terms of algebraic formulas. On the side of phenomenology, he is best known for "I think, therefore I am." Thinking cannot be characterized in terms of extension in physical space. Thus, an essential distinction between mental and physical is established. The awareness of thoughts and thinking, as in Descartes' dictum, is just one category of experience that is described in phenomenology, and a statement that better captures phenomenology in its full generality is "I sense, therefore I am."

> When, on a common-sense basis, people talk of the gulf between mind and matter, what they really have in mind is the gulf between a visual or tactual percept and a "thought" —e.g., a memory, a pleasure, or a volition. But this, as we have seen, is a division within the mental world; the percept is as mental as the "thought". (HK, 228)

Since I have spent two chapters framing the dualism of mind and body in terms of phenomenology and physics, I simply credit Descartes for clearly delineating these two realms of study, and proceed directly to two philosophers who attempted to circumvent the problematic dualism of mind and body that Descartes had formulated.

Bishop Berkeley proposed a world comprised only of human minds and the mind of God. In this conception, God coordinates our phenomenal perceptions in just such a way that the presumed evidence for a physical world is accounted for without need for the physical world itself. The gist of this can be conveyed in terms of God as a hypnotist. If a hypnotist could plant appropriate perceptions directly into the minds of an audience, he could make the audience see an elephant, and see an elephant disappear, without need of a real elephant. The elephant represents the physical world. If the sense impressions that we take to indicate the physical world are planted in our minds directly by God's will, then the existence of the physical world is superfluous. Berkeley's conception thus yields a complete elimination of all things physical.

That the natural world should be a mere show put on for the benefit of human minds strains the credulity of most of those minds, and Berkeley fails to convince us. However, Berkeley's demarcation between self-evident phenomena and the inferred physical world survives today in the understanding that conjecture is inherent in all scientific knowledge and that belief in physical entities amounts to a provisional hypothesis. The same principle, in a religious context, is called "faith."

Leibniz gives us a significant variation on Berkeley's theory in his *Monadology*. I will take liberties with this view, extracting what seems pertinent to an eventual solution of the problem. Leibniz invites us to consider a swarm of fish, perceived from a distance. It might be mistaken for a lifeless mass. Look closer, and we find a multiplicity of living individuals. Use a microscope to examine one of the fish more closely and we find a multiplicity of living cells. We now know that the process of "looking closer" comes to an end at the quantum level. Any hopes that some residue of matter would be yielded by the investigation also come to an end. The natural world consists therefore of immaterial entities. Though this seems strange, there does not seem to be any argument about it. Leibniz supposed, even without the benefit of quantum theory, that the human percipient, when investigating the natural world, is following a trail of perceptions that leads to... other percipients! A percipient, or perceiving subject, Leibniz calls "a monad." A human mind is an example of a monad. The natural world is a system of monads. Most monads are presumably less sophisticated than human monads, but each has its own sensory experience. Monads are the ultimate individuals, the mentalistic "atoms" that replace the material particles of a physicalistic conception of the world. Descartes' dualism is overcome, as with Berkeley, by restricting the world to mental experiences. In contrast to Berkeley, Leibniz affirms our intuition that the natural world consists of something beyond our own minds and perceptions—namely, the monads, which have *their* own minds and perceptions. The natural world is thus an environment that is teeming with minds and nothing else.

As we have discussed, the very notion of physical is bound up in the notion of physical space and what it contains. Leibniz avoids the assumption of physical space as a "container" for his monads. He does not conceive the monads to be *in space*, but instead finds the ordering principle of space *in the monads*. Like Berkeley, Leibniz accounts for the locations of physical space by reference to the perspectives inherent in the phenomenal visual fields of the individual monads. That is, we normally explain a visual perspective as being due to a location and orientation in physical space. Berkeley and Leibniz invert this, reconstruing physical space as a correlation of phenomenal visual fields. One can conceive an ordering, of the phenomenal fields-of-view belonging to individual monads. Each monad is endowed with its own perspective view. Thus, the monads are ordered among themselves by perspective variations of their internal visual experiences, without invoking either physical space or causal interaction among the monads.

Any thoughtful attempt to reinterpret the meaning of physical space is important for the mind-body problem, and we should note the strong and weak points of Leibniz' interpretation. The virtue of his hypothesis is that space is defined in terms of the same geometric features that we are acquainted with in our visual experience. This avoids the need to postulate a physical space that transcends experience, which would then have to be somehow "tied back" to our sensory experience in order to justify the postulate with empirical consequences. We have noted that belief in material substance was such a postulate that is now discredited. Leibniz' theory of space can be appreciated as an early attempt at an "operational definition" of space in phenomenological terms. Operational definitions are employed in order to nail down abstract concepts to the matter-of-fact sequence of steps that one takes when conducting experiments to demonstrate a theory. This approach minimizes questionable theoretical assumptions. When I measure my bedroom for a carpet, I go through the experience of handling and viewing a measuring stick, marking the landing of its endpoint, moving it along in steps to traverse the room, and keeping count of the steps. Leibniz takes this series of experiences, and others like it, to *be* the dimensions of my room. The supposedly physical nature of these dimensions consists in the fact that anyone who bothered to repeat my experiences would come up with the same count of steps as I do.

The weakness of Leibniz' theory is that it gives no indication why the experience of one monad should bear any correlation whatsoever to the experience of another. The unifying space in his theory is a pure effect, without any significance as a causal factor. Space is just an outcome of all the perceptual experiences that all the monads happen to have. Leibniz therefore appeals to a "pre-established harmony," worked out when God determined what the experiences of each monad would be. Once created, the monads are strictly isolated in the privacy of their individual experiences and they do not interact with one another. The spirit of science has been just the opposite, to find the patterns of causal interaction between the parts of the world. In this spirit, science defines a spatial location for every entity, and this location is a causal factor in every interaction.

We are likely to dismiss the theories of Berkeley and Leibniz as curious attempts to deny a physical world of insentient stuff, since we have come to accept the latter without qualms. How could these acclaimed geniuses stray so far from commonsense in their beliefs? The answer is that commonsense is thoroughly infected with irreconcilable beliefs in minds and bodies, making it the truly curious theory, if it can even be called a theory. I don't mean to assert that the theories of Berkeley or Leibniz are correct. But they *are* logically coherent possibilities, which is more than one can say for commonsense dualism. It is worth the imaginative effort required to suspend disbelief and conjure up vividly what each man, Berkeley and Leibniz, proposed. On the one hand, it helps to glimpse the extent to which phenomenological experience accounts for the world when physical entities are left out. On the other hand, the meaning of "physical entity" can appear in sharper relief by subtracting the purely phenomenal ingredients of the world employed by Berkeley and Leibniz from the hybrid sum of phenomenal-and-physical that comprises the commonsense view.

I will venture to describe the commonsense view of mind and body. Bodies are stuff without any mentality or feeling whatsoever. Some bodies though, are alive, and do have mentality and feeling—human bodies specifically. Our subjective experience has a role in determining the physical behavior of our bodies, providing us with the means to control our actions. At the same time, our bodies and brains determine to a great extent, if not completely, what we experience. This two-way determination is so smooth and seamless that no sharp distinction can be drawn between mind and body.

I will comment on the preceding from the point of view taken in this book, starting with the last sentence. The claim that "no sharp distinction can be drawn between mind and body" is denied. The first two chapters are intended to draw just such a distinction. Phenomenology and physics have come apart, and if this dissociation is likened to a divorce, physics was the party that filed for separation. The sensory data of mental experience was evicted for being incompatible with the entities championed by physics. Science has a de-anthropomorphizing effect with its mode of explaining things, and people have a vague uneasiness that this mode of explanation does not stop short at human behavior. It does not stop short. Science thoroughly excludes human sentience in principle, as explained in the previous chapter. No one is comfortable with this and it is rude to draw attention to it.

Regarding the causal interaction between mind and body, as held by commonsense, this will be vindicated in subsequent chapters that deal with the solution. The smooth and seamless nature of this interaction will acquire a natural explanation. But this will demand a revised explanation of physical space, which is beyond the current horizon of common sense, and beyond the goal of this chapter. In the current context, minds and bodies cannot interact because body-to-body interaction is the only kind of interaction that physics deals with. Bodies collide with other bodies, not with colors. As bodies have now been reinterpreted as probability waves, we have instead that probabilities are calculated against other probabilities, not against colors. In any case, the entities of physics interact with others of their own kind, not with the kind we know as sensory qualities. Thus, physics does not support causal interaction between sentient experience and the body, and it ignores sentient experience altogether. Physics and commonsense therefore diverge on this point, which should be the source of some anxiety.

Due to the same considerations, physics does not support the commonsense idea that living bodies have feeling or sentience. Rather, physics conceives a living body to be a complex variant of the same entities that account for the non-living world. Trying to bring intention, volition, or feeling into the definition of biological life is called "vitalism," which is treated as superstition by contemporary science. By the end of the book, we shall concur with commonsense that living bodies have feeling, but only as part of the general conclusion that all bodies, living and non-living, consist ultimately of sentient occasions of feeling.

I heard an interview on the radio with a woman who works on artificial intelligence at MIT. This woman also has a background in theology, which she finds relevant to her work with two computerized robots in the MIT laboratory. These two robots learn new responses through interaction with humans. The woman had begun to bond emotionally with the pair of laboratory creations, and she was nearly ready to attribute *personhood* to them. She also referred to the android character Data, from the Star Trek series. Data is my favorite character on that show, so I was pleased to find that the woman took an interest in him. Completely missing from the discussion was the question of whether Data, or the MIT robots, might have any feeling, or sentience. Therefore, consideration was confined to the intelligent aspects of mind, which Herbert Feigl has termed "sapience" to distinguish philosophical problems pertaining to intelligence from problems pertaining to sentience. I was amazed that theology could be brought to bear on the budding personhood of a robot without raising the issue of feeling. While I do not doubt that computerized modeling of human behavior is interesting and important, it is confined to physical mechanisms, the common ground for analyzing human behavior and computer behavior. How theology could be thought to pertain to beings without pleasure, pain, or sense of self, I have no idea.

If Data does have feelings, then we have ethical issues as to how he is treated. It is feelings, and only feelings, that can open a discussion to matters of values, morals, rights, or theology. A great many science fiction stories sidestep the issue of whether an android has feelings. You can infer that an android has a sensorium when a movie shows you what the android sees. You never seem to hear an android's internal monologue, even though verbal narration is commonly employed to convey the inner thoughts of human characters. You just see what the android sees. Other than that, you must interpret the android's facial expressions, bodily actions, and what it says about itself in order to judge whether it has sentient experience. Generally, an android is impervious to pain, and displays an affect that indicates minimal emotion. Optional program modules are sometimes installed to outfit an android with further capabilities. Insofar as these upgrades confer further physical behavior capabilities, implementation via software upgrade is perfectly understandable. However, that digital programming should endow the android with some type of phenomenological experience makes no sense at all.

Nevertheless, we are entertained by the ambiguous personhood of Data and his kind. I think that this indicates that people today do have a latent interest in the mind-body problem. The android is a near-human, with questionable inner life. He simulates a person whose vitality is at low ebb, a person who is just "going through the motions." We root for Data and his inner life because it is increasingly necessary for us, in the mechanistic age of science, to root for ourselves.

...

I am about finished trying to impart the mind-body problem to the reader. I was talking to someone recently about the problem. In response to my contention that qualitative color is not part of the scientific account of the physical world, he said, "The brain just *interprets* physical stimuli as colors." Well-satisfied with that, he exited the conversation. His solution made light work of philosophy. I wondered later whether he would be equally happy with a reverse formulation: "The mind just *interprets* colors as physical stimuli." The latter is more in accord with philosophy of science. Most people are barely aware of analytic philosophy and the difficulty in framing a systematic view of the world. At the same time, philosophy has no satisfactory view of the world to teach, or so it is widely believed. Hence, our educational system imparts knowledge of specialized fields without any overall coherence. This lack of overview is not acknowledged. It would be nice if a high school education culminated in a coherent view of the world, according to which the graduate could choose a future role in society. When it is suggested that one's view of the world is not coherent, one feels a personal affront to one's rationality. The affront though, is to mankind in general, as the history of philosophy shows.

The mind's habitation of the body has been likened to a "ghost in a machine." The ghost, which has all the experience, goes undetected by science, which deals only with the machine. This is another way of acknowledging the fracture of the world into irreconcilable halves—what Whitehead called the "bifurcation of nature." My avenue to the mind-body problem was Whitehead's *Science and the Modern World*, which might serve the reader where I have failed. It provides an account of the history of human thought in respect to scientific developments, with the mind-body problem serving as the pervasive connecting thread.

CHAPTER 4

....

Relations and Structure

Relations account for whatever order and structure are to be found in any realm of investigation. Relations and structure are among the phenomena presented to our sentient minds. Relations and structure form the basis of mathematics, and together with causal assumptions, the basis of physics.

In the present chapter we shall be concerned with a purely logical discussion which is essential as a preliminary to any further steps in the interpretation of science. The logical concept which I shall endeavor to explain is that of "structure". (HK, 250)

Russell, in the context of his own book, is leading up to a new definition of "physical." The meaning of the terms "structure" and "relation" must be well established in order to grasp further important definitions. For instance, an event is to be classified as "physical" if it has *causal relations* to other events, and space-time is asserted to be the *causal structure* of events. The terms "causal," "relation," and "structure" will have to shoulder an appreciable load of meaning, since they will be used to replace, and invalidate, the notion of "physical" given in Chapter 2.

> We can now proceed to the formal definition of "structure." It is to be observed that structure always involves relations: a mere class, as such, has no structure. Out of the terms of a given class many structures can be made, just as many different sorts of houses can be made out of a given heap of bricks. Every relation has what is called a "field," which consists of all the terms that have the relation to something or to which something has the relation. Thus the field of "parent" is the class of parents and children, and the field of "husband" is the class of husbands and wives. Such relations have two terms, and are called "dyadic." There are also relations of three terms, such as jealousy and "between"; these are called "triadic." ... To this series of kinds of relation there is no theoretical limit.

> Let us in the first instance confine ourselves to dyadic relations. We shall say that a class *alpha* ordered by the relation R has the same structure as a class *beta* ordered by the relation S, if to every term in *alpha* some one term in *beta* corresponds, and vice versa, and if when two terms in *alpha* have the relation R, then the corresponding terms in *beta* have the relation S, and vice versa. ... (HK, 254)

Let's consider an example that illustrates the definition of "same structure." A class is just some definite set of entities. Let us define the class *alpha* to have as members two streets in my neighborhood, Central and Lowry. Let us choose *intersect* as a relation between streets. We can then write *intersect* (Central, Lowry) to state that Central and Lowry intersect, which is true. Next, let us define the class *beta* to have as members my two children, Aly and Andy. We'll choose *sibling* as a relation between children, and write *sibling* (Aly, Andy). We now set up a one-to-one correspondence between the members of *alpha* and the members of *beta*, such that Central corresponds to Aly, and Lowry to Andy. Under this correspondence, when two terms in *alpha* have the relation *intersect*, then the corresponding terms in *beta* have the relation *sibling*, and vice versa. Therefore, the two classes, ordered by their respective relations, have the same structure.

The two relations, *intersect* and *sibling*, are each symmetrical. That is, if Central and Lowry intersect, then Lowry and Central intersect. Consider the asymmetrical relation, *parent*. We define another class *gamma* with members Carey and Aly. If *parent* (Carey, Aly) is true, then *parent* (Aly, Carey) is false. Now let us test whether the class *alpha* ordered by *intersect* has the same structure as the class *gamma* ordered by *parent*. We set up a correspondence of Carey to Central, and Aly to Lowry. Due to the symmetry of the *intersect* relation, we have true statements in both *intersect* (Central, Lowry) and *intersect*(Lowry, Central). The correlative to the latter statement, pertaining to the class *gamma*, is *parent* (Aly, Carey), which is false. The test for same structure has failed. In general, a class ordered by a symmetrical relation does not have the same structure as a class ordered by an asymmetrical relation.

Suppose we augment the class of streets to contain a third member, Johnson, which intersects Lowry but runs parallel to Central. We'll also augment the class of children to include my nephew, Wil. Because Wil is sibling to neither Aly nor Andy, but each street intersects with at least one of the other two, we find that no matter how we set up a one-to-one correspondence between the members of the two classes, the test for same structure fails.

The relations and classes we have chosen to compare do not lead to structural uniformities of any mathematical interest or importance. Nevertheless, the examples illustrate the definition of structure in terms of relations and relata. (The individuals connected by a relation are called its "relata.") Furthermore, we have identified common structure in facts as disparate as intersecting streets and sibling-related children. Any fact includes a logical structure of relation and relata exhibited in that fact but exhibited also in facts belonging to other realms of discourse. Of special relevance to the analysis of mind and body, common structure can be identified between facts of phenomenology and facts of physical science. In the commonsense view of interaction between mind and body, we distinguish objects in the physical environment by virtue of a display of colored patches in our visual experience. This coordination of mind and body can be specified according to structure that is common to both realms, even though each realm, as characterized in previous chapters, is composed of its own proprietary relations and relata that are absent from the other realm.

If we use "R" as a variable that stands for *any* relation, and if we use "x" and "y" as variables that stand for *any* relata, we get an expression like "R(x, y)," which shows the logical form of a class of rudimentary facts. Pure mathematics is concerned with such expressions. A "dyadic" relation connects one individual to one other individual. If dyadic relation R is such that R(x, y) and R(y, z) implies R(x, z), then R is said to be a "transitive" relation. If R is dyadic, asymmetrical, and transitive, then all the individuals that it relates to a given individual will form the type of structure called a "series." Serial structure can apply to geometric entities, such as points which form a line, or to non-geometric entities, such as auditory notes ordered by higher-in-pitch to form a scale. The mathematical characterization of a relation, such as "dyadic," "asymmetrical," or "transitive" determines the variety of structure that can be formed by that relation. Two relations with the same mathematical properties can form the same structure out of dissimilar relata. In other words, common structure can be identified across two separate realms of entities which otherwise have nothing in common.

> When two complexes have the same structure, every statement about the one, in so far as it depends only on structure, has a corresponding statement about the other, true if the first was true, and false if the first was false. ... (HK, 255-256)

Note Russell's use of the word "complexes" above. A complex is a "whole," and a whole is not just the logical sum of its parts.

For example, you can't build a computer from a parts list alone. You need a schematic to describe how the parts fit together. The schematic shows the structural arrangement of the parts. In ordinary usage, the word "structure" has two meanings that could produce confusion in the current discussion. On the one hand, a house is commonly referred to as "a structure." On the other hand, in line with the meaning of *relational* "structure," a house made of bricks may have the same structure as another house made of stones. Using both meanings of "structure" together, "the two structures have the same structure," which highlights the potential confusion. I will confine the use of "structure" to Russell's definition, so that any complex (or whole, or fact) has relational structure which specifies how the primitive relations and relata fit together to form a complex. With this usage, *relations*, *relata*, *structure*, and *complex* are four nouns with distinct meanings.

Pure mathematics has more general concerns than our actual world. It is concerned instead with the description of all possible structures that arise from all possible types of relation. This unlimited prospect could degenerate into a tedious catalogue of trivial variations, so it is constrained by intuitive criteria of elegance, beauty, and power. When some field of pure mathematics finds application in scientific theory, the physical world itself seems to manifest the elegance of the mathematics. To *apply* mathematics to the real world, one must provide names for specific relations and relata thought to belong to the real world, and these names are substituted for the variables in mathematical expressions in order to make logically coherent statements about the world. In this naming and substitution, the logical distinction between relations and relata must be systematically obeyed.

> Let us take next the relation of a district to a map of it. If the district is small, so that the curvature of the earth can be neglected, the principle is simple: east and west are represented

> by right and left, north and south by up and down, and all distances are reduced in the same proportion. It follows that from every statement about the map you can infer one about the district, and vice versa. ... These inferences are possible owing to identity of structure between the map and the district.

> Now take a somewhat more complicated illustration: the relation of a gramophone record to the music that it plays. It is obvious that it could not produce this music unless there were a certain identity of structure between it and the music, which can be exhibited by translating sound relations into space relations, or vice versa; e.g., what is nearer to the center on the record corresponds to what is later in time in the music. It is only because of the identity of structure that the record is able to cause the music. ... (HK, 253)

In the first paragraph above, identity of structure is found in a map and a district, both physical entities. In the paragraph following it, the gramophone record is a physical entity, but "music" and "sound" could be interpreted as either physical patterns or phenomenological patterns. There is common structure with the gramophone record in either case.

> ... A wireless set transforms electromagnetic waves into sound waves; a human organism transforms sound waves into auditory sensations. The electromagnetic waves and the sound waves have a certain similarity of structure, and so (we may assume) have the sound waves and the auditory sensations. Whenever one complex structure causes another, there must be much the same structure in the cause and in the effect, as in the case of the gramophone record and the music. This is plausible if we accept the maxim "Same cause, same effect" and its consequence, "Different effects, different causes." If this principle is regarded as valid, we can infer from a complex sensation or series of sensations the structure of its physical cause, but nothing more, except that relations of neighborhood must be preserved; i.e., neighboring causes have neighboring effects. ... (HK, 254)

Russell is bringing cause-and-effect into the discussion of structure, and specifically, physical causes of sensory experience. That is plausible until we consider that science has distilled its knowledge of cause-and-effect into the laws of physics. As ordinarily understood, the laws of physics pertain to quantitative energy components in the geometry of space-time, with no intelligible link to the sensory qualities of experience. Accordingly, similarity of structure between sensory experience and physical events constitutes a coincidence rather than a causal connection. This affront to common sense will be redressed in the next few chapters, where the causal interaction of mind and body is rescued, but at the expense of the conventional interpretation of physics.

> Take, for example, the question of waves versus particles. Until recently it was thought that this was a substantial question: light must consist either of waves or of little packets called photons. It was regarded as unquestionable that matter consisted of particles. But at last it was found that the equations were the same if both matter and light consisted of particles, or if both consisted of waves. Not only were the equations the same, but all the verifiable consequences were the same. Either hypothesis, therefore, is equally legitimate, and neither can be regarded as having a superior claim to truth. The reason is that the physical world can have the same structure, and the same relation to experience, on the one hypothesis as on the other.

> Considerations derived from the importance of structure show that our knowledge, especially in physics, is much more abstract and much more infected with logic than it used to seem. There is, however, a very definite limit to the process of turning physics into logic and mathematics; it is set by the fact that physics is an empirical science, depending for its credibility upon relations to our perceptive experiences.... (HK, 256)

Russell and Whitehead, as a pair, are best known for their joint creation, *Principia Mathematica,* 1910-13. This work presents a systematic construction of mathematics in a formal system of symbols. The system made do with a tiny vocabulary of symbols representing logical notions, such as if-then, and, or, not, any, identity, some, all, set, and set membership. The effect was to collapse formal logic and mathematics into a single system of expression and calculation, erasing the distinction between the two fields of study.

It is with these credentials that Russell explains the fundamental role of relations in defining structure and mathematics. Russell and Whitehead did not collaborate explicitly after *Principia Mathematica*. Each of them turned next to the philosophy of science. By 1927 Russell had published *The Analysis of Matter*, and Whitehead had delivered the Harvard lectures which were published two years later as *Process and Reality*. To my mind, the agreement of ideas between these two books furnishes the solution to the mind-body problem, which surpasses their achievement in the foundations of mathematics. It seems likely that they became uniquely equipped for their later insight by their earlier collaboration, particularly, by fastening upon relations as the key to logical analysis.

While belief in relations as a fundamental sort of entity is especially strong in Russell and Whitehead, it is especially weak in most of us. We're quick to affirm the existence of objects, substances, qualities, and even physical space, as "things." But we're reluctant to think of relations among things as further things. We would rather think of relations as projected somewhat arbitrarily by thought onto the intended objects of thought or perception. A non-committal attitude about relations serves to shield us from the obligation to analyze physical space into the sort of relations that distinguish physical space from phenomenological spaces, or from purely mathematical spaces. This may explain why Russell and Whitehead's solution has not been widely recognized and heralded by the academic community.

Belief in relations can be fortified through two considerations. One is the fruitfulness of naming relations and making use of these names. As Russell contends, this yields a general understanding of pure mathematics, as well as the application of mathematics to the real world. Another consideration that reinforces belief in relations is that sensory experience is partially constituted by relations that we directly perceive. To take a visual example, three colored patches spaced apart in a row present a self-evident type of "between" relation. Secondly, for a non-geometric type of between-ness relation, orange is between red and yellow (in the hue circle) with respect to the relation of color similarity. In phenomenology, in science, or in any field requiring logical description, there is no fruitful account of order or structure without the straightforward acknowledgment of relations as irreducible components. This sets the stage for an examination of space, time, and causality in terms of "causal relations."

CHAPTER 5

Space-time as Causal Structure

Special Relativity eliminates instantaneous spatial relations in favor of time-ordering causal relations. Causal relations are definable without recourse to geometric notions. Time order, for physics, is relative position in a causal chain of events. Two events not ordered by a causal chain are called "contemporaries." Spatial order is defined for contemporaries by the convergence of their respective causal chains at common causal ancestors and descendants.

In this chapter, we wish to apply the understanding of relations and structure covered in the previous chapter to the analysis of physical space. The usual understanding of physical space is confined to geometric features such as areas, volumes, points and lines. These features can be ascribed to a person's visual experience, even during dreaming. Science must have a space that is consistent for all observers, a harmonization of perspectives based on waking perception, discarding the data of imagination and dreams. Furthermore, the space of science must do without color or sensory qualities in its definition. It follows that we do not *perceive* physical space. We are restricted to *conceiving* it. The same holds true for the entities which populate physical space, since the whole apparatus of physics is refined from the commonsense belief in a world not limited to our sensory experiences. Though we might be conscientious about this distinction between perceived space and the space of physical theory, we unhesitatingly borrow the geometric features of visual space and carry them over to the space of physics, as we do when learning Euclidean geometry in school. The problem then is to distinguish pure geometry, which is mathematics, from the geometry of our actual world, which is physics. We shall find the requisite distinction in the very purpose of science, which is to build a predictive causal framework from our scattered perceptions.

> ... The brain is in the head, but thoughts are not—so, at least, philosophers assure us. This point of view is due to a confusion between different meanings of the word "space." Among the things that I see at a given moment there are spatial relations which are a part of my percepts; if percepts are "mental," as I should contend, then spatial relations which are ingredients of percepts are also "mental." Naïve realism identifies my percepts with physical things; it assumes that the sun of the astronomers is what I see. This involves identifying the spatial relations of my percepts with those of physical things. Many people retain this aspect of naïve realism although they have rejected all the rest.

> But this identification is indefensible. The spatial relations of physics hold between electrons, protons, neutrons, etc., which we do not perceive; the spatial relations of visual percepts hold between things that we do perceive, and in the last analysis between colored patches. ... (HK, 201-202)

> ... When I am said to 'see' a table, what really happens is that I have a complex sensation which is, in certain respects, similar in structure to the physical table. The physical table, consisting

> of electrons, positrons, and neutrons, is inferred, and so is the space in which it is located. It has long been a commonplace in philosophy that the physical table does not have the qualities of the sensational table: it has no color, it is not warm or cold in the sense in which we know warmth and cold by experience, it is not hard or soft if "hard" and "soft" mean qualities given in tactile sensations, and so on. All this, I say, has long been a commonplace, but it has a consequence that has not been adequately recognized: that the space in which the physical table is located must also be different from the space that we know by experience. (HK, 221-222)

The mind-body problem would not have been solved without the discovery of a limiting velocity in the universe. Without this knowledge, which required centuries of scientific progress, belief in an extended space enduring through time could not be seriously challenged. In retrospect, this absolves classical philosophy from its failure to solve the mind-body problem, and the furthering of science was the inadvertent but essential step to promoting a solution. It is well known that three-dimensional space and one-dimensional time give way in Special Relativity to four-dimensional spacetime. This hyphenation of "space" and "time" is due to Einstein, and we wish to understand in this chapter how causal relations emerge as the basis of spacetime order.

> ... Physical space is wholly inferential, and it is constructed by means of causal laws. Physics starts with a manifold of events, some of which can be collected into series by physical laws; for example, the successive events constituting the arrival of a light ray at successive places are bound together by the laws of the propagation of light. In such cases we use the denial of action at a distance not as a physical principle but as a means of *defining* spacetime order. That is to say, if two events are connected by a causal law, so that one is an effect of the other, any third event which is a cause of the one and an effect of the other is to be placed between the two in spacetime order. (HK, 222-223)

Using Arrows to Illustrate Causal Relations

Let us use an arrow as a graphic element to represent a causal relation. The direction of the arrow indicates the asymmetry between cause and effect. The arrow will be the only graphic element. It is left for the reader to imagine that every arrow implicitly connects a causal event at the tail of the arrow to a causal effect at the head.

A Causal Relation Between Two Events

For the time being, let us consider causes and effects to be simply whatever kind of entities can consistently be understood to have causal relations to one another. We will use the term "event" by which to refer to the primitive causes and primitive effects at the bottom layer of causal analysis. We imagine therefore a causally primitive event at each end of an arrow.

We could join the head of one arrow to the tail of another. The undepicted event at the junction of these two arrows is an effect with respect to one arrow and a cause with respect to the other. Using letter-names for the implicit events, we have a situation depicted in which A causes B, which in turn causes C. This represents a minimal causal chain of events.

A Causal Chain of Events

Next, consider two arrows joined at their tails, suggesting a forking path. This represents one event that has unmediated causal influence upon two others. In this case, the common causal event is called the "causal predecessor" of the other two events, and these latter events are called the "causal successors" of the first event.

One Event with Two Causal Successors

The other primitive formation is two arrows which meet at their heads, representing two events which have a combined and unmediated causal influence upon a third event. Here we have two events that are the causal predecessors of a single causal successor.

Two Events with One Common Causal Successor

We can now begin to imagine elaborate drawings of any desired complexity, using arrows that fork at their tails, and arrows that meet at their heads. We shall add another naming convention pertaining to arrow diagrams. If an event "Z" (that is, some particular junction of arrows) can be reached from another event "A" by tracing a path that consistently obeys the direction of arrows, then Z is a "causal descendant" of A, and A is a "causal ancestor" of Z. No path of arrows shall be drawn, which followed in the direction of its arrows, completes a circuit. This ensures that no event shall be its own causal ancestor or its own causal descendant.

We shall conflate causal order and time order, so that our graphs depict "the arrows of time." Cause-and-effect order agrees strictly with time order. Causal relations thus have the generic character of before-and-after relations. There is only one kind of temporal succession, which is the same as causal succession. Accordingly, an event can have more than one temporal predecessor and more than one temporal successor, as shown in the two previous diagrams. That will allow us to graphically construct the 4-D manifold and the common particles from temporal succession, showing that discrete time is the only parameter required for the theory of physics.

An arrow drawing is a graphic aid to conceiving causal relations and causal structure. The two-dimensional page contributes some geometry that is irrelevant to what is being represented. For instance, the length of arrows and the angles they form at the junctions are irrelevant. Only the order of connection of the arrows is relevant.

We can form a new diagram from the two previous diagrams, which we shall call "a primitive diamond."

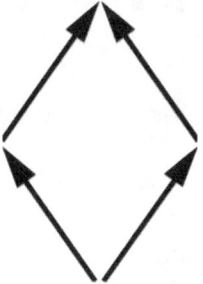

The Primitive Diamond

We can replicate one primitive diamond across the width of a canvas to form a horizontal row of diamonds abutting at their left and right corners. We can then replicate the entire row up and down the canvas, such that the canvas becomes covered with a perfectly monotonous diamond pattern. This uniform pattern of arrows could represent a two-dimensional spacetime. Each interior event is at the junction of two arrows arriving and two arrows departing.

THE MIND-BODY PROBLEM AND ITS SOLUTION

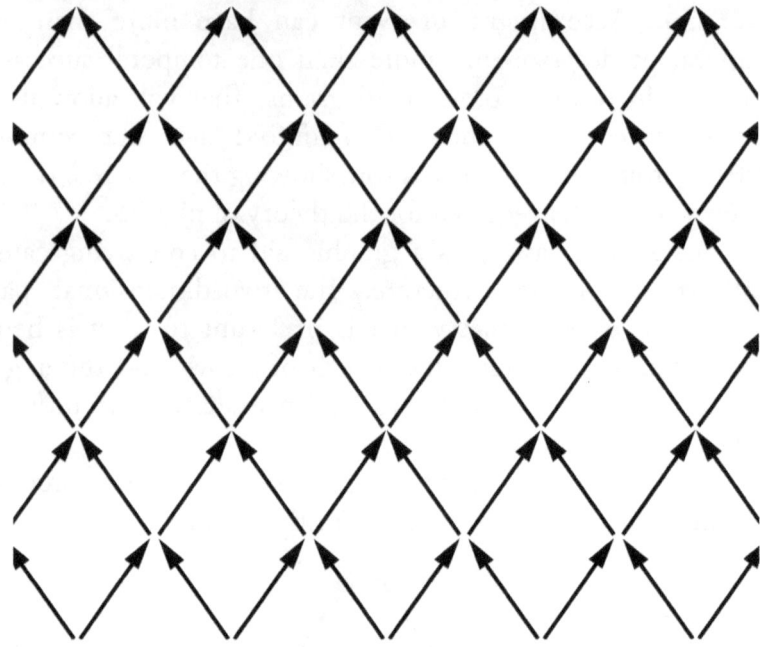

A Uniform Diamond Pattern

Consider an event "E" in the uniform diamond pattern. Starting at this event, a path can be traced, obeying the direction of arrows, to arrive at various other events. The set of events that can be reached starting from E is the set of its causal descendants-- "the future of E." Conversely, "the past of E" is the set of E's causal antecedents.

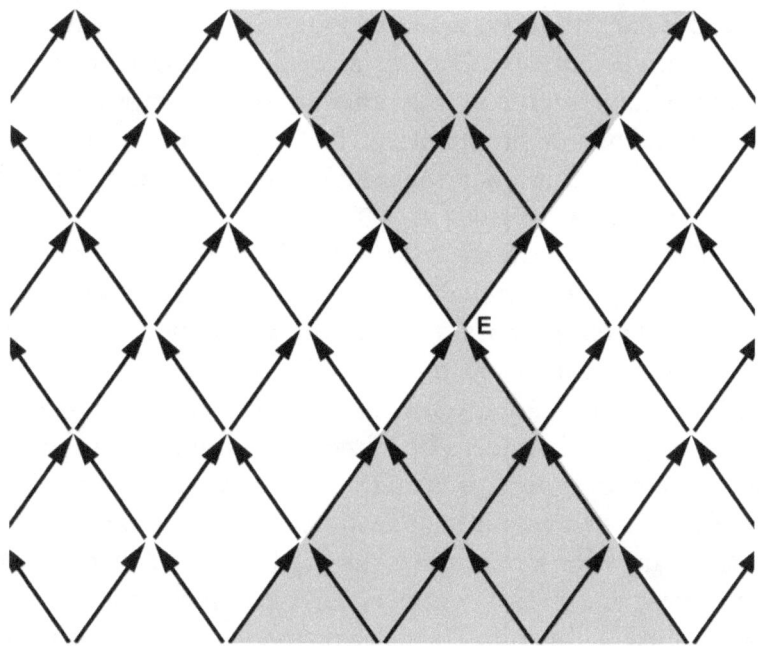

The Causal Past and Causal Future of E

When the past and future of E are filled in with shading, there remain unshaded regions. These regions contain events called "the contemporaries of E." While the contemporaries of E belong neither to E's past nor to its future, they do not, contrary to our usual intuition of time, form a class of events simultaneous with E. There are many causal chains lying wholly within the region of E's contemporaries, and these causal chains imply the passage of time, which is incompatible with simultaneity. Before

Special Relativity, it was assumed that, with respect to any given momentary event, simultaneity must define a unique spatially extended universe. The abandonment of this assumption is called "the breakdown of simultaneity." It is very difficult to let go of the belief that there is a momentary "now" spatially extended throughout the universe.

Let us designate as "causal chain" any path of arrows that can be traced by obeying the direction of the arrows. In that case, the contemporaries of E are the events not connected to E by any one causal chain. In Special Relativity, causal chains are called "world lines," and the relation between two events on a world line is said to be "time-like." The relation between two contemporary events is said to be "spacelike." The textbooks usually depict three out of four dimensions of space-time (two spatial dimensions, one dimension of time.) In that case, the causal past and causal future of an event form two opposing cones called "light cones." Finally, textbooks illustrate spacetime as continuously divisible, which is convenient mathematically for describing large aggregates of events.

> The continuity of space-time, which is technically assumed in physics, has nothing in its favor except technical convenience. It may be that the number of space-time points is finite, and that space-time has a granular structure, like a heap of sand. Provided the structure is fine enough, there will be no observable phenomenon to show that there is not continuity. (HK, 291)

> We will show that there is *not* continuity by obtaining quantum theory automatically from the assumption that time is the discrete next-to-next succession of moments. The step of time will be identified as the quantum. Our time diagrams will then constitute quantum schematics, with each arrow depicting a quantum.

If we want a diagram of Newton's space and time, for comparison with Special Relativity, we need two graphic elements, one for spatial relations and another for time relations. An arrow serves as a time relation, indicating the asymmetry of before-and-after. A short line segment, without an arrowhead, showing no asymmetry of direction, represents a spatial relation. Satisfied again with depicting only two dimensions, a dashed line stretching horizontally represents a one-dimensional line of space at one instant of time. Each line segment explicitly represents the relation of spatial contiguity between a point of space at one end and another point of space at the other. The spatial relations form a *line of simultaneity* slicing across the universe. Another such horizontal dashed line placed above the first represents space at the next moment. A vertical arrow drawn from the lower line to the upper line indicates the time-ordering relation. In this conception of space and time, the moments of time form a single series. This means that arrows do not form forking paths as they do in the Relativity diagram. Instead, all arrows line up head-to-tail in single-file.

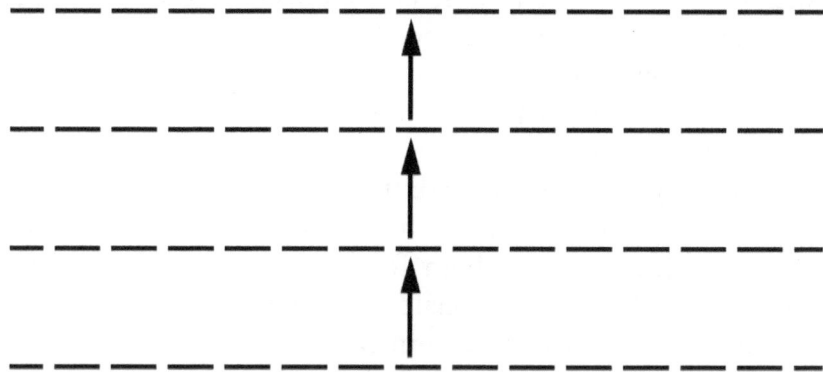

Newtonian Space and Time Relations

In the Newtonian diagram, we cannot, as we can in the Relativity diagram, interpret the arrows to be causal relations adequate for defining both space and time. Newton's physics located each entity by its spatial location plus its temporal location. The units of measure for space and time were incommensurable.

Space relations and time relations were distinct types of relation used to define causal order. Space and time were conceived independently before a theory of cause-and-effect was framed.

We are now in position to grasp the import of Special Relativity for our notions of time and space, and the diagrams help make this clear. The uniform diamond pattern represents a spacetime of causally related events, the arrows representing causal relations. Paths that obey the direction of arrows represent causal chains. Events connected by a causal chain have, in the terminology of Special Relativity, time-like relations to one another. Contemporary events, which have spacelike relations to one another, are connected only by pathways that do *not* obey the direction of arrows. The crucial insight to be gained is that we have no need for spatial relations as a primitive type. We will retain the term "spacelike" for the separation of contemporaries because their separation is entirely due to the lack of true spatial relations. True spatial relations form lines, surfaces, and volumes without any reference to time or the direction of time. We are eliminating true spatial relations from physics in favor of time relations alone. We can dispense with the term "time-like" because our arrows depict real time transitions, which are not merely *like* time-- they *are* time.

We begin to see that, with the advent of Special Relativity, physics can be built up from causal relations, or time relations, alone. I will quote from both Russell and Whitehead in this regard, since the point is crucial for re-interpreting the character of physical space, which in turn proves crucial to solving the mind-body problem.

> We think, for example, that it is possible to move from A to B or from B to A; but such a view is incompatible with the theory of spacetime. According to that theory, every position of a body has a date, and it is impossible to occupy the same position at another date, since the date is one of the co-ordinates of the position. When we travel from A to B, the date is continually advancing; the return journey, having different dates, does not

> cover the same route. Thus geometry and causation become inextricably intertwined.

> ... Dr A. A. Robb has laid stress upon the fact that, when two events have a spacelike interval, there can be no direct causal relation between them. This means that, given two such events A and B, if any inference is possible from the one to the other, it must be by way of a common causal ancestor. ..." (AM, 313-314)

> It is the definition of contemporary events that they happen in causal independence of each other. Thus two contemporary occasions are such that neither belongs to the past of the other. The two occasions are not in any direct relation to efficient causation. The vast causal independence of contemporary occasions is the preservative of the elbow-room within the Universe. ... Nature does provide a field for independent activities. ... (AI, 195)

In the following, Whitehead uses the word "occasion" where Russell would use "event." Also, a "nexus" is any definite set of connected occasions.

> The notion of the contiguity of occasions is important. Two occasions, which are not contemporary, are contiguous in time when there is no occasion which is antecedent to one of them and subsequent to the other. A purely temporal nexus of occasions is continuous when, with the exception of the earliest and the latest occasions, each occasion is contiguous with an earlier occasion and a later occasion. The nexus will then form an unbroken thread in temporal or serial order. (AI 202, 203)

The above quote describes a one-dimensional series of occasions. With the phrase "temporal or serial order" Whitehead is equating temporal order and serial order. That is very conventional, since time is usually conceived as strictly one-dimensional. However, we shall depart from that convention in the use of the terms "time and "temporal." We are eliminating spatial relations from physics altogether, and we shall ascribe the term "temporal order" to any set of causally connected occasions, linear or not. It is clear from the graphs that the time-ordering relation depicted by the arrow is the *only* ordering relation employed in our reconstruction of physics.

In the remainder of this chapter, we shall concentrate on time diagrams and their interpretation. We shall account for the common particles as patterns of time sequence. A limiting velocity for the motion of bodies is a simple consequence of this analysis. Special Relativity is usually presented as the various consequences of a limiting velocity, with the velocity limit serving as a brute axiom. Distortions are then ascribed to space and time, which vary with the frame of reference used for making measurements. Formulas are then obtained to incorporate the limiting velocity (the speed of light) into all calculations of velocity and energy. But the distortions of space and time are just as mysterious as the brute axiom of a velocity limit which the distortions are designed to accommodate. By contrast, the limiting velocity is a simple consequence of the reduction of space-time to discrete time. *Any change in space-like separation is a purely structural consequence of the stepping advance of time.* Thus, "space cannot outrun time."

The explanation for the limiting velocity is the original impetus for the reduction of spacetime to a causal network of time-ordered moments. The next two chapters will provide additional support for that reduction. Firstly, the reduction of physics to time-ordered moments makes intelligible the location of mental events in the physical world. Secondly, we shall follow Russell's reasoning that the scientific method can at best discover the *causal skeleton* of the world.

An arrow connecting two events represents the causal influence of one event upon the other, while the absence of an arrow connecting two events implies the lack of causal influence of either event upon the other. Just as an event either happens or it doesn't, we are supposing that any given event either has a direct causal effect upon another given event or it doesn't. That corresponds to an arrow being drawn or not. In Newtonian physics, causal influence shaded off as a function of spatial distance without ever quite reaching zero. In Special Relativity however, events with spacelike separation have no causal influence at all upon one another. This supports an all-or-none analysis regarding causal relations.

The primitive diamond represents a well-defined case of causal structure. The hollow diamond is easy to pick out visually wherever it occurs in a diagram. If quantum events are the primitive events of physics, susceptible to no further causal analysis, then it takes two or more quantum events causally connected to one another to form a pattern of activity that *does* have a definitive causal analysis. If such a pattern of activity is repeated along a causal route, we are apt to call this "a particle." In the causal analysis of spacetime, there are no material particles as fundamental entities. There is only the relentless cause-and-effect succession of immaterial events. Particles and bodies are causal patterns that recur in this process.

We have been considering the time order of events. Now we shall consider the measurement of a time period, or duration. Look at the following primitive diamond that also includes a vertical arrow drawn directly from bottom to top.

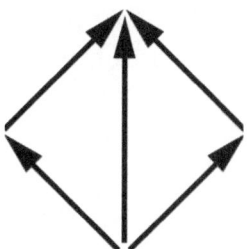

Frequency Ratio 2:1

There are three causal routes from the bottom event to the top one, and the middle route looks like a "short cut." Either of the side routes involves two causal transitions with an intermediate event along the way. All three routes determine the same temporal interval, since they all begin at the same time and end at the same time. We can apply the term "frequency" in comparing these alternative routes that have the same origin and destination. Either side route has twice the temporal frequency of the middle route.

The next diagram of alternate routes between common end points has two arrows forming one route and three arrows forming another, producing a temporal frequency ratio of 2:3.

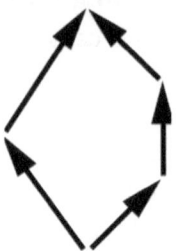

Frequency Ratio 2:3

Diagrams exist for frequency ratios ranging through the rational numbers. Thus, we can assign numerical measure to temporal frequency ratios. The recognition that time can form frequency ratios is very simple. It could have been discovered by anyone. Yet it eluded Russell and Whitehead and it continues to elude the physics world. The fact that time can form frequency ratios, without recourse to any sort of matter-in-motion, is the key to reducing physics to time as the sole parameter.

Quantum theory was born when Max Planck discovered that atoms only emit light at discrete energy levels. The energy levels are proportional to the frequency values of the emitted light. Planck expressed the coupling of energy and frequency with the formula $E=hf$. The constant of proportionality, "h," is known as Planck's constant. We have seen that discrete time can form frequency ratios. By Planck's formula, two energy values, E_1 and E_2, are equal to two frequency values, hf_1 and hf_2. Thus, $E_1 / E_2 = f_1 / f_2$. (Planck's constant cancels out.) This gives us the opportunity to *define* energy ratios as frequency ratios. The step of time is the unit of the frequency ratios. Thus. the step of time is the quantum of the energy ratios. We have the definition of energy ratios and their quantum in terms of time alone. Energy is nothing else than the stepping action of time. The single step of time is the quantum of energy.

The reciprocal of a temporal frequency is a measure of duration—the measure of a time period or time interval. For example, a frequency of 4 steps of time per second implies a duration of ¼ second for each step of time. Higher frequency sequences consist of shorter duration steps. The steps of time are formed in a full range of frequency ratios, with reciprocal ratios of duration. In this theory, which lacks true spatial relations with their own metric of spatial intervals, duration will serve not only as the measure of time but also as the measure of spacelike separation.

...

After I published the first edition of this book in 2004, I extracted the diagrams for physics into a booklet, "A Theory of Everything for Physics." I started out with a systematic survey of the simplest time diagrams. There are four valid time diagrams that connect exactly three moments.

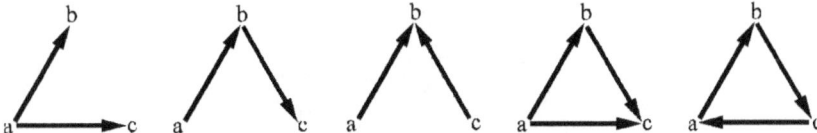

Time diagrams with 3 moments

From left to right, we can recognize fork, series, and convergence. The fourth diagram is the simplest to show frequency ratios. It has two pathways from "a" to "c." One path takes two steps while the other path takes one step, forming a frequency ratio of 2:1.

The diagram on the far right is not a valid time diagram. I show it for that reason. Each moment in that diagram is its own causal ancestor, which is prohibited. No moment can be earlier or later than itself.

Next, I constructed all the valid diagrams that have exactly 4 moments. When I came to the graphs that have 5 moments, there were too many. I narrowed my survey to highly symmetrical graphs. I came to the following graph of 6 moments connected by 10 arrows, which I shall call the "hex cell."

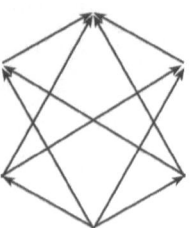

The Hex Cell

By using the hexagon shape, we can "stencil" six hex cells around a center one, arriving at an extendable four-dimensional time lattice.

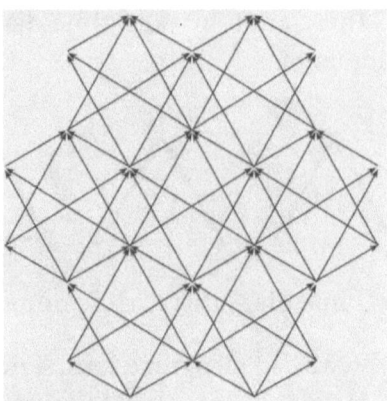

Extendable 4-D Time Lattice

Each interior node is at the intersection of four arrows arriving and four departing, which is the hallmark of four-dimensionality.

The number 137, which is called "the fine structure constant," is used extensively in physics to perform calculations. It has had, up until now, no physical interpretation. Richard Feynman called it "the greatest damn mystery in physics." The following diagram has 137 arrows.

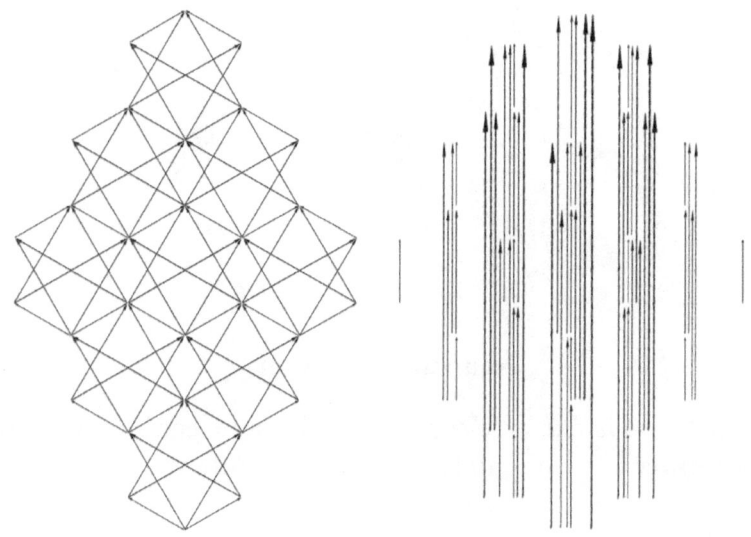

137 Arrows

The 59 vertical arrows, displaced to the right, show all the arrows that could connect pairs of vertically aligned moments of the 78 4-D lattice arrows on the left. To have found such a symmetrical graph of 137 arrows made me think that "closed diagrams" are extremely important. A closed diagram has a single earliest moment and a single latest moment. Such a diagram can be used in chained repetition to produce a particle-like sequence persisting in time. The following hex cell formations are closed diagrams In chained repetition, they depict forms of neutrino propagation:

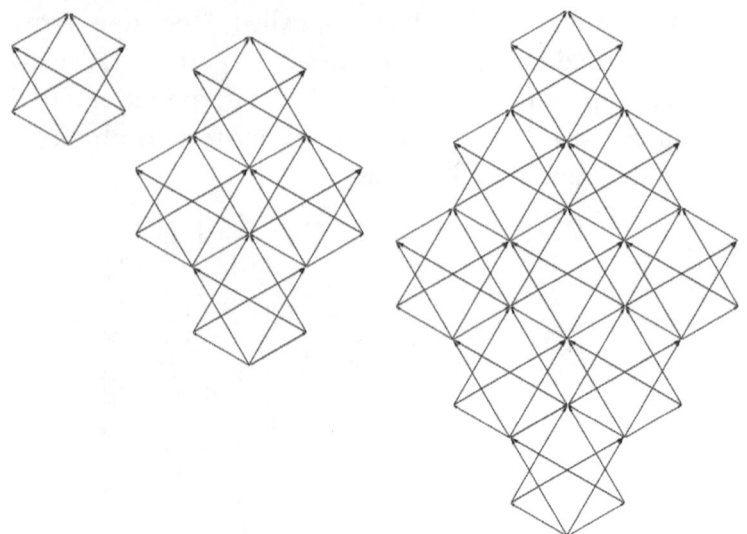

Neutrino formations

The neutrino is like an electron but without electric charge. The hex cell can accommodate further quanta of charge and momentum.

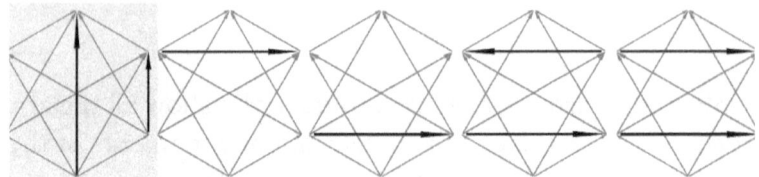

Quanta of charge and momentum

The vertical arrows represent quanta of forward momentum. The horizontal arrows, which necessarily break the bi-lateral symmetry of the cell, represent charge quanta. With the presence of charge quanta, the hex cell serves as an electron cycle, which propagates in chained repetition to form an electron. The following set of diagrams, if charge quanta were shown, depict, from left to right, a free electron, two of its atomic cloud formations, and an encounter with a photon.

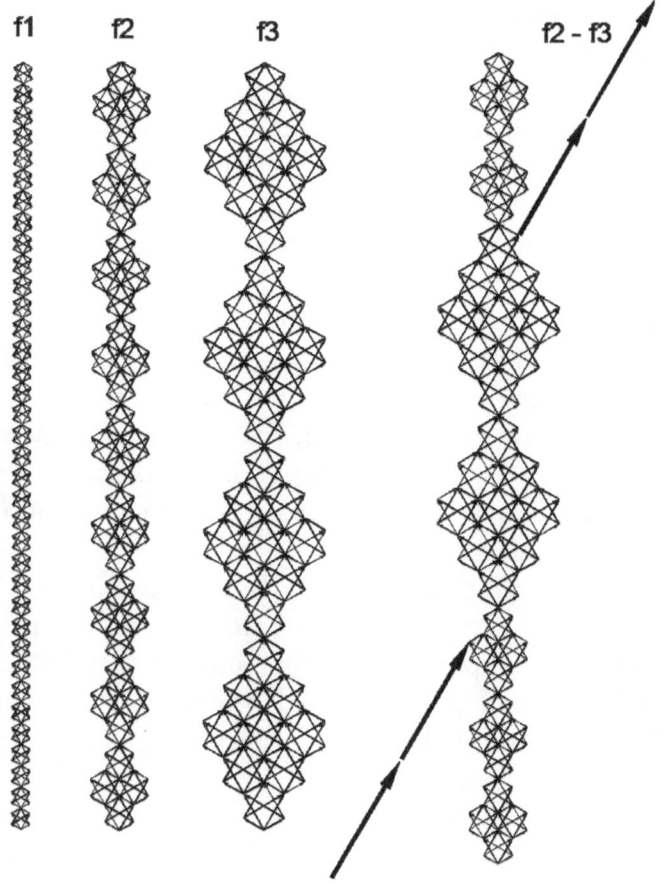

Graphical Account of Bohr's Formula

We have, left to right, an electron, a hydrogen cloud, a helium cloud, and lastly, a hydrogen cloud disturbed by an incoming photon which is later emitted. Each sequence is formed by chained repetition of its unique characteristic cycle. Each such cycle has its frequency of chained repetition, labeled at the top as f1, f2, and f3. Each sequence has 36 hex cell components. Also, each sequence is scaled to reach the same height on the page as every other sequence. If we take the hex cell as the unit of energy, then the scaling shows equal amounts of energy transpiring in equal amounts of time.

Bohr's formula gives the main spectral frequency values of light emitted by the atomic electron clouds. Bohr's formula for wavelengths is R x $(1/k^2 - 1/n^2)$. (Ignore the Rydberg constant "R" for the moment.) If we take the wavelength of a free electron as 1 and its frequency as 1, then the wavelength of the hydrogen cloud is 4 and its frequency 1/4; the wavelength of the helium cloud is 9 and its frequency 1/9. Thus, the frequency of the photon depicted in the diagram (the inverse of its wavelength) is 1/4 − 1/9, satisfying Bohr's formula for wavelength. The corresponding formula for frequency is f2 − f3. The graphic patterns extend to produce Bohr's formula along the entire atomic series, explaining the discrete frequency values of light emitted and absorbed by the atomic electron clouds.

The constant value "R" in Bohr's formula was ignored. It is a product of other constants that are each reducible to 1. Our electron cell provides a natural unit for mass and charge—two of the components of the Rydberg constant. Then there is Planck's constant "h," which cancelled out when we defined energy ratios as frequency ratios. Finally, the absolute speed of light "c" is commonly considered a scale factor for converting seconds to meters and is set to 1. That is especially appropriate in the reduction to time, where time measure is employed as the measure pf space-like separation. We are left with the diagrams themselves, from which all the number and structure of physics derives. The free electron serves as the "real-time clock" that sets the base frequency for the electro-magnetic spectrum.

> The electron cycle is a hex cell, from which the 4-D manifold is also formed. Notice the redundancy among the 4-D manifold, the atomic cloud formations, and the forms of neutrino propagation, which are all composed of hex cells. An economical hypothesis is that the electron clouds and the neutrino formations together constitute all that there is of the 4-D manifold. The manifold is made of quanta, and it need

not be as uniform as usually assumed. Gaps in the manifold delineate the locally separate electron clouds and neutrino formations. That would explain why neutrinos seem so elusive to the physicists. They are part of spacetime itself.

The account of Bohr's formula promises other fruitful avenues of investigation. The inverse squares in Bohr's formula are tied to the structure of electron clouds that comprise part of the 4-D manifold. Thus, the inverse squares in Bohr's formula are likely tied to the inverse square laws for electricity and gravity. Secondly, the stepped-up scaling of clouds in the atomic sequence depicts a case of *discrete time dilation*. Time dilation is an aspect of the "curvature of space-time" in Einstein's theory of gravity, the General Theory of Relativity. The discrete time dilation in the account of Bohr's formula can serve as the basis for a discrete version of General Relativity. Solving the problem of quantum gravity is a top priority for physicists today.

Nuclear Structure—the 6-D Time Lattice

Attached to periodic nodes of an electron cloud sequence is a nuclear sequence comprised of higher frequency quanta. This constitutes a synchronization of the nucleus to its electron cloud, which assigns the nucleus its location in the 4-D manifold. The "attachment" of nucleus-to-cloud can only mean that the two discriminable sequences share periodic nodes.

The three magnetic strut models above are all assembled according to a single lattice principle. This lattice principle has widespread practical use in building design, for which Buckminster Fuller named it "the octet truss." The assembly in the middle can be described as a unit octahedron (of edge length 1) with a tetrahedron erected on each face. On the right we see a cuboctahedron, with 6 square faces and 8 triangle faces. "Caged inside the cuboctahedron" is a center ball-bearing with 12 radial struts connecting to the 12 outer vertices. If we erect a pyramid on each square face of the cuboctahedron, we get an octahedron of edge length 2, as shown on the left. (Its bottom pyramid is omitted in order to allow the assembly to stand upright.) The octet truss is extendable in all directions, so that the above forms will recur periodically in the lattice. Larger versions will be formed, at twice the size, three times the size, and so on.

The spatial lattice exemplified by the models becomes a time lattice when each strut is assigned an unambiguous direction, while obeying the directionality of time. We thereby obtain a richer structure composed entirely of time-directed pathways, with each strut depicting a quantum. When converting the spatial lattice to a time lattice, the conversion can be performed in such a way as to maximize the preservation of symmetry. Also, the conversion can be performed by choosing different directions through the spatial lattice. For example, the direction can be chosen such that the octahedron on the left has a first moment at the bottom and a last moment at the top. Alternatively, the conversion of the star-shaped form in the middle can be performed such that one outer point becomes the earliest moment and the diametrically opposed outer point becomes the last moment. The octahedron on the left, at three times the size shown, will, in chained repetition, depict a neutron.

The star-shaped model in the middle has its outer points spaced at the corners of a virtual cube. Thus, that model can be assembled with others like itself in a 2x2x2 arrangement to make the next larger cube. The next larger cube is formed with a 3x3x3 arrangement. It has the perfected symmetry and number of quanta to model, in chained repetition, a proton.

With the cuboctahedron on the right, we have 12 struts radiating from the center point. After conversion to a time lattice, there will be 6 arrows arriving at the center and 6 arrows departing. As the lattice is extended, every interior point will have that same context—the defining feature of 6 dimensions. The 4-D lattice structure of the electron clouds is distinct from the 6-D lattice structure of a nucleus, but the two types of lattice have hexagonal components in common. The octahedron model has six outer hexagons encircling the assembly like great circle routes on a globe. Such hexagonally spaced nodes are present on the outer surface of the proton and neutron models as well. The hex cells of an electron cloud can share nodes with the virtual hexagons on the surface of a nucleus, completing those lowest frequency hexagons with the hex cell quanta of the companion electron cloud. For more detail about the nucleus, and calculation of the mass-ratio of proton and neutron with respect to the electron, see the following paper online.

"Causal Set Theory and the Origin of Mass-ratio"

Quantum theory is reconstructed using standalone causal sets. The frequency ratios inherent in causal sets are used to define energy-ratios, implicating the causal link as the quantum of action. Space-time and its particle-like sequences are then constructed from causal links. A 4-D time-lattice structure is defined and then used to model neutrinos and electron clouds, which together constitute a 4-D manifold. A 6-D time-lattice is used to model the nucleons. The integration of the nucleus with its electron cloud affords calculation of the mass-ratio of the proton (or the neutron) with respect to the electron. Arrow diagrams, along with several ball-and-stick models, are used to streamline the presentation.

http://vixra.org/pdf/1006.0070v1.pdf

(Today's causal set theory is the same as Whitehead's discrete event ontology in respect to the formations that can arise from sheer temporal/causal succession.)

The constructions above explain what particles are, what mass is, and why particles have the masses they do. Particles are repetitious patterns of time sequence, and the mass-energy of a particle is the measure of its inherent relative frequencies, which is a function of the number and arrangement of its constituent quanta. A particle sequence must in turn have a frequency relative to other particles, since all frequency is relative. All quanta of this universe are causally connected. That is what "this universe" means.

The standard expectation today is to find randomness and chaos at the bottom of things. To explore random arrangements of causal links is to expect the greatest complexity of nature at its basis. Instead, the simplest, most symmetrical patterns of causal links quickly reveal the most fundamental entities and features of physics, which can be no accident. In retrospect, reliance on randomness at the foundation of physics is the most hazardous assumption, though it is commonly considered to be the minimal assumption and the safest.

Because of the discovery of the frequency ratios, it appears that the "Big Bang" has been misconstrued as a singularity marking the beginning of time when it is more likely that our system of temporal succession has no beginning. As we consider earlier and earlier stages of the universe, we conceive a spatially shrinking universe, converging quite naturally according to our spatial intuitions to a point-like minimum of spatial extent. But the clarified situation involves higher frequencies, and the "shrinking of space" is a pure consequence of these increasingly higher frequencies. An issue of Discover magazine featured a marble-sized sphere on the cover graphic that represents the spatial size of the universe at $t=10e-34$ seconds. Yes, space shrinks toward zero as all frequencies increase, but clarity on the subject again demands that we relinquish unanalyzed spatial intuitions. The negative exponent in the equation is heading toward larger integers (higher frequencies) as we delve into earlier stages of the universe. Nothing in the basis of our theory suggests that frequencies should have an upper limit, unless we restrict consideration to a bounded region. Nothing suggests that the exponent in the equation will, with further knowledge on our part, attain some satisfactorily high integer and reach a "glass ceiling." Physicists of the distant future, should they be likewise captive to their spatial intuitions, might well conclude that our age was one of unbearable heat, a marble-sized universe suffering from excruciating proximity to a cataclysmic Big Bang.

...

It follows from the above constructions that spacetime order, measurable intervals, quantifiable energy, particles and their masses, can all be defined as structural features arising from causal relations among events. We have devoted no consideration to the intrinsic nature of events themselves, or to the intrinsic nature of causal relations. The events and transitions of time constitute "what happens" in the universe, but what are they? That question is addressed in the remaining chapters. In this chapter, we just consider events and causal relations to be deliberately hypothesized entities that provide a framework for modern physics with a minimum of assumptions.

In the light of the foregoing, let us reconsider the notion of the physical world described in Chapter 2. I said there that the meaning of "physical" amounts to an intuition of space and what's in it. Special Relativity implies that determinate location in space-time is due to time-ordering relations alone, as indicated by the arrow diagrams. The theory abandons purely spatial relations that define space as instantaneous extension. Since one event can have several immediate causal predecessors or successors, we must conceive the course of time as branched into locally separated streams. For physics, spacelike structure is part and parcel of this richer structure of time. The notion of space as an extended state is abandoned. This demands a corresponding revision to our conception of physical entities as inherently spatial. That conception belongs to a provisional stage of science that has been overturned by the discovery of a limiting velocity. The analysis of the universe into whole-and-part now finds the parts to be immaterial events that occur in somewhat regular patterns of succession. These patterns account for physical space, energy, shape, location and motion. This constitutes the de-materialization of matter and the dismantling of physical space required by Special Relativity. The meaning of "physical" therefore devolves upon events, their time relations, and the resulting patterns.

The patterns resulting from the succession of events are variations of structure. The types of structure that can arise are due to the logically definable attributes of the time relation. The time relation is, by hypothesis, asymmetrical, irreflexive (no event is its own ancestor), and sufficiently multi-termed to relate several individuals asymmetrically to several others. But these logical attributes define a general relation for pure mathematics, not a relation that has specially to do with time and events. The mathematical expressions of physics can only pertain to the actual world if the time variable in those expressions refers to actual events and facts about their causal succession. The physical nature of defined structures, such as space, energy, and particles, derives entirely from the presumed physical nature of their structural ingredients-- namely, events and the causal pairing relations that link events together.

With the understanding that causal succession is equivalent to temporal succession, physics amounts to a theory of what comes before what. Progress in this theory has arrived at quantum events—discrete events which do not admit of further before-and-after analysis. These rudimentary events, and the time relation that orders them, are the quintessential physical entities. Our impression of the physical world as something substantial and immense owes to the sheer number of quantum events and their causal connections. If the subject of scientific investigation is simply the temporal sequence of events, then the physical world is not well characterized as "space and what's in it," but rather "time and what's in it." This has immediate consequence for the mind-body problem, since there is little difficulty in assigning the time of occurrence to a mental event.

CHAPTER 6

....

The Physical Location of Mental Events

Mental events have physical location by the same criterion as physical events, strictly by the theory of their causes and effects. Mental events are between their causes and effects, and this causal positioning is the complete criterion and meaning of their physical location, as it is for events in general, mental or non-mental.

The previous chapter suggests that physical theory can be put into canonical form in terms of the relation of cause-and-effect and its relata. It is evident that the causal relation, symbolized by the arrow, is the fertile element providing the variations of structure, while the class of event-like relata is the logical residue remaining after causal structure is worked out. The events have the logical status of individuals required for the causal relation, but beyond that, it is difficult to know what to make of them. The pattern of their succession accounts for time, space, energy, and all things physical. Yet a single causally primitive event, considered in isolation, has no physical properties whatsoever. The temporal *transitions* are the quanta. They have the frequency/energy values, not the events. The positing of quantum events allows us to conceive the physical world as patterns of concatenated quanta, while the intrinsic nature of the momentary events never becomes pertinent to physics. Physics has nothing to say about what a quantum event might be.

The intuition of spatialized existence has served for ages to help us conceive the physical world, from "earth, air, fire and water" to tinker toy molecules and planetary electrons. Any portion of spatial existence is again spatial existence, right down to the imagined point of space. The causal analysis of Chapter 5 construes any fact about space to be a logical consequence of facts about time. It would contravene the analysis to revert to the model of spatialized existence to help conceive quantum events, because all space-like relations are derived solely from the relative placement of these events in time. Thus, Descartes' conception of physical existence as characterized by spatial extension is undermined by the causal analysis of space-time. This deprives us of a traditional attribute of physical entities that could serve to differentiate physical events from mental events.

We're still left with a useful distinction between mental and physical events—namely, that we know nothing of the intrinsic nature of physical events, but quite a lot about the intrinsic nature of mental events. When we attend to the time order within our experience, we glimpse the elusive fleeting moment of experience. There is more to describe about this moment of experience than how elusive it is. For example, consider the visual image presented by this page of text. The stable presence of black markings against a white background endures throughout a continuous stretch of many present moments. In a phenomenological description, the visual field presents us with a striking two-dimensional expanse, accompanied by a somewhat feeble dimension of depth. This geometry of the visual field does not flicker or fade as we narrow attention to a minimal time slice of experience. Rather, the space of the visual field pervades the present moment and is a part of it. From the standpoint of phenomenology, there are spatial relations that order the distinct regions of visual space into an ordered whole that is presented all-at-once, in a single moment of experience. This is in marked contrast to the space-time of physics, which is rid of instantaneous spatial relations. This highlights the difference between the visual space of experience and the causal space-time of physics, which can help us avoid mistaking one for the other.

Not only does a mental event provide a full-fledged visual space, it provides further ingredients not subject to the spatial relations of the visual field, such as itches, odors, and sounds. Let us define a "mental event" explicitly as "the full cluster of phenomena, culled from all the senses and modes of awareness, that co-exist simultaneously in one moment of experience." A person's enduring sentient experience can then be considered a temporal succession of such mental events. This prepares us to examine the time correlation of mental events and physical events.

As we examine time relations between mental events and physical events, it is difficult to avoid the topic of causal interaction between the two. The main objection against such causal interaction is two-fold. First, mental events are essentially composed of sensory qualities, while physical events are distilled by causal analysis from a scientific legacy of theoretical entities not thought to possess any sensory qualities. Thus, mental events and physical events would seem to be so dissimilar as to make their causal interaction unintelligible. Secondly, physical entities such as particles and electromagnetic fields are all "carved out of space," co-defined with respect to one another to give them coherent roles of interaction, while mental events are not defined with any spatial shape that would allow them to be congruent with physical entities. The latter objection loses its force when causal analysis is carried through to completion, since we then arrive at events that have no inherent spatial properties. That leaves only the first objection, which represents a reluctance to introduce sensory qualities into physics only to satisfy the rare occurrence of mental phenomena in a predominantly non-mental universe.

> But although physics as a self-contained logical system does not need to mention sensations, it is only through sensations that physics can be *verified*. It is an empirical law that light of a certain wavelength causes a visual sensation of a certain kind, and it is only when such laws are added to those of physics that the total becomes a verifiable system. (HK, 261)

Russell seems to suggest a two-tiered theory to forestall the objection to introducing sensory qualities into physical theory. We have a skeletal theory of cause-and-effect, depicted by arrow diagrams, which describes the physical world without mention of sensory qualities. But this theory provides no means of verification for the sentient observer, who observes nothing but sensory qualities. Therefore, we could form an auxiliary theory that includes mental events, in order to account for the verification of the skeletal theory by sensory observation. The auxiliary theory incorporates the skeletal theory in its entirety and adds the postulate that physical events have causal influence upon mental events, or at least the class of mental events that Russell calls "percepts." This allows us to describe the role of sensory observation in experimental science, while maintaining a clear distinction between mental events and "physics as a self-contained logical system."

> The theory that perceiving depends upon a chain of physical causation is apt to be supplemented by a belief that to every state of the brain a certain state of the mind "corresponds," and vice versa, so that given either the state of the brain or the state of the mind, the other could be inferred by a person who sufficiently understood the correspondence. If it is held that there is no causal interaction between mind and brain, this is merely a new form of the pre-established harmony. But if causation is regarded—as it usually is by empiricists—as nothing but invariable sequence or concomitance, then the supposed correspondence of brain and mind tautologically involves causal interaction. The whole

> question of the dependence of mind on body or body on mind has been involved in quite needless obscurity owing to the emotions involved. The facts are quite plain. Certain observable occurrences are commonly called "physical," certain others "mental"; sometimes "physical" occurrences appear as causes of "mental" ones, sometimes vice versa. A blow causes me to feel pain; a volition causes me to move my arm. There is no reason to question either of these causal connections, or at any rate no reason which does not apply to all causal connections equally. (HK, 196-197)

We might try to express the notion of time without presupposing causality by saying "events just happen." But events happen with some predictability, which leads us to believe that past events influence future events. Thus, time order is enmeshed with causal order. The arrow diagrams of Chapter 5 are interpreted as representing either time order or causal order, if one wishes to maintain a distinction between the two. In any case, the order of succession of events accounts for the spacelike order of contemporary events in space-time. Since mental events and physical events are interspersed in a common time ordering, the spatial location of mental events with respect to physical events is assured. Under Special Relativity, the facts of time sequence determine spacelike relations, regardless of whether the time-ordered events are mental, physical, or a mix of the two.

Now we may consider the experimental means by which the space-time location of human mental events is determined. It will come as no great surprise to find them located in the vicinity of the human head. The method is called a psycho-physical experiment. I shall be the imaginary subject of such experiments, for the purpose of discussion. I submit to having my skull cap removed and micro-electrodes planted in my cortex. An electroencephalogram, or EEG, records evidence of a concert of electrical activity at the surface of my cortex. There are electrodes for sensing electrical activity, and other electrodes for inducing electrical currents. In the first experiment, an electrode is used to stimulate a pinpointed region of my visual cortex. I see a flash of light in my visual field. Whether my eyes are open or closed, I see a flash of light whenever the stimulus is applied. The electrical event apparently causes me to see the flash. I then report that the flashes I am seeing are moving progressively to the left in my visual field, which elicits a murmur of approval among the scientists. It turns out that an electrode was being progressively repositioned along a line running across my visual cortex. The spatial geometry of my visual field can be "mapped out" to correspond to the surface region of my brain called the "primary visual cortical projection area."

Next, a series of stimuli is applied to some spot on my visual cortex at a rate of one per second. I see a series of flashes occurring at this same rate. The rate is gradually increased, until I report that the intermittent flashes have fused into a steady, stable point of light occupying my visual field continuously. This fusion occurs when the rate of excitation exceeds 10-per-second.

Suddenly I hear an unexpected sound. The electrode has been applied to a point on my auditory cortex. After further eliciting of sounds, and sound sequences elicited at varying rates, the stimulus phase of this experimental session is over.

For the final experiment, I am to practice meditation. I close my eyes, calm my thoughts, and I reach a relaxed but alert state of awareness. I remain in this condition for a few minutes. Wires are then unhooked, my skull cap replaced, and I'm shown the EEG record produced during my meditation period. I have produced a nice train of *alpha wave patterns* on the EEG recording. These are synchronous oscillations at a steady rate of ten-per-second.

Consider first the clues offered by the experiment regarding the time period of a mental event. The fusion of light flashes into a steady spot in my visual field occurred when the electrical stimuli surpassed a rate of ten-per-second. The same thing happened with sounds. Experiments suggest that a human mental event, as we have defined it from a purely phenomenological point of view, can be assigned a time period of roughly one tenth of a second. We cannot discriminate stimuli that occur at a higher rate than this. There is therefore no point in supposing that it takes a succession of more than ten mental states to account for a person's sentient awareness over the span of one second. We do not experience more *changes of state* than ten-per-second. It is unwarranted to ascribe any finer discrimination of time than this to human awareness. Secondly, a meditating subject can *reach alpha*, a phenomenological condition of calm alertness corresponding to a pronounced synchronization of EEG oscillations at a rate of ten-per-second. Since the concert of brainwave activity recorded by an EEG is the closest known physiological correlate to a person's mental state, a rate of ten-per-second for human mental events has further experimental support. Without evidence to the contrary, this seems like a reasonable number if *any* temporal rate is to be assigned to the succession of human mental events. Periodicities in the human brain likely account for the characteristic form of perceptual experience shared by humans. This calls for a compatible periodic rate of mental events, rather than a haphazard rate, or no rate at all. We shall therefore proceed under the assumption that experience transpires in "drops" of roughly one tenth of a second duration.

A rate of 10-per-second for human moments is also appropriate to the delays involved in the conduction of efferent nerve signals from the brain to the muscles, and in the reverse direction, the conduction of afferent signals from the sense receptors to the brain. *Reaction time*, to avert a driving collision for example, is not reducible to less than one tenth of a second. Reliable motor control of the body requires patience for the feedback, which is subject to the propagation delays of neural transmission. The human series is well qualified for central control of the human body, equipped at 10 Hz with the ideal frequency for the job.

Strobe lights at 10 Hz bother people, and epileptics are prone to seizure when they see such strobe lights. All in all, given that we are seeking a finite frequency for the human series, a regular frequency of 10 Hz seems to be it. We are not aware of this frequency by introspection. It is ascertained only in the laboratory, by reference to scientific hypotheses concerning a world that lies beyond the reach of anyone's introspective powers.

Our sentient experience of time seems to be smooth and without breaks. The best we can do to account for that smoothness, using discrete time analysis, is to model the human series as an unbroken alternation of moment, transition, moment, transition. We then have a typical discrete time series, constituted by moments and transitional quanta, which we shall term "human moments" and "human quanta." Thus, a human series consists of human moments connected by human quanta.

Putting a number to a moment of human experience, with dimensions in seconds, establishes a commensurability of mental events and physical events. Regarding the physical location of mental events, we have the electrical events induced by probes at the surface of the cortex as the closest known causal predecessors to effects in the sensorium. Unless there are further psycho-physical experiments that can indicate an even closer time relationship between physical excitatory events and what I see and hear and feel, we have mental events "sandwiched" between known physical events to the highest precision afforded by experimental means. The electrical excitations at my cortex which elicited the visual and auditory sensations in my experience are the physical events contiguous in time to my mental events.

Let's confine discussion for the moment to the sense of vision. With eyes open, light from the surrounding environment is focused by the lens to fall upon the retina, exciting the rods and cones in a pattern that reproduces a scene from the surrounding environment similarly to the way film is exposed in a camera. We may call this pattern of excitation on the retina a "virtual image." The excitations at the retina result in the propagation of neural signals along routes through the optic nerve bundle. If someone snipped my optic nerves, the scene would be lost from my experienced visual field. With optic nerves intact, the virtual image at the retina is transmitted to the optical cortex to be reproduced there. This reproduction relies mainly on two features of the nerve signal propagation. First, individual pathways constitute "point-to-point wiring," connecting individual rods and cones of the retina to individual sites on the visual cortex. Secondly, a greater intensity of excitation at an individual rod or cone results in a higher frequency of nerve cell discharges along the respective path of transmission. Thus, light intensity is "encoded" in the frequency of nerve cell discharge, and spatial patterns are preserved by point-to-point wiring of retinal sites to sites on the optical cortex. Therefore, when I am viewing a painting, I am enjoying its color and composition due to the timing and spatial arrangement of electrical discharges of nerve cells at my optical cortex. The time course of events leading up to activity at my cortex is once-removed, or twice-removed, or even more remote, from the external causes of my visual experiences.

What is known about the other human senses, such as sound and touch, involves similar stories of cause-and-effect, with nerve pathways and the frequency coding of nerve cell discharges leading to activity levels at the cortical surface of the brain as the closest concomitant in time to the features experienced in a mental event. There is nothing controversial in this causal theory of perception, which is of a piece with the scientific account of physical events in general, whether inside or outside the human body. The point to be emphasized here is that the order of cause-and-effect, when applied to the mixed domain of physical events and mental events, while situating mental events earlier and later than certain physical events, pertains to a specific class of physical events that are spatially located at the cortex of the brain. Thus, a human mental event, by being adjacent in time to a class of physical events at the cortical surface of the brain, has its causal location narrowed by psycho-physical experiment to that brain location.

From the viewpoint adopted in this chapter, in which physical events and mental events are kept distinct, human mental events might seem to be strange interlopers among the tissues and electro-chemical events of the science and physiology of the brain. We tend to think that the inter-relations of physical parts of the brain are open to inspection, and that the physical location of any of these parts with respect to others is a matter of direct perception. By contrast, the mental events interspersed with brain activity are only privately experienced by the subject of the experiments. It is only by interpreting the subject's verbal reports that the experimenter can infer the occurrence of mental events, which can then be correlated with physical events. Thus, it seems that mental events have only "second-hand" physical location, borrowed from the concomitant physical events of the brain, while these latter events establish physical location in the primary sense.

While it is true that inference is required to attribute mental events to the human subject of an experiment, similar inference is required to attribute a brain to this same subject. The experimenter presumably has his or her own brain, which is also subject to the causal theory of perception, even when it is not the focus of an experiment. According to physical theory, the experimenter infers the entire physical environment, including the subject's brain tissues, from events in his own cortex. Furthermore, if the experimenter is like you or me, his primary data does not seem to come in the form of cortical events at all, but rather in the form of qualitative sensory data, such as colored patches, sounds, touch, and pressure. At this point, our presumed familiarity with physical objects is threatened, and we're likely to take an abrupt defensive tactic. The burden of a train of inferences involved in the perception of physical objects is thrown out the window, and percepts in the mind of the physiologist are mistaken for the brain that is being experimented upon.

> Then, again, there is the argument about brain and mind. When a physiologist examines a brain, he does not see thoughts; therefore, the brain is one thing and the mind which thinks is another. The fallacy in this argument consists in supposing that a man can see matter. Not even the ablest physiologist can perform this feat. His percept when he looks at a brain is an event in his own mind, and it has only a causal connection with the brain that he fancies he is seeing. When, in a powerful telescope, he sees a tiny luminous dot, and interprets it as a vast nebula existing a million years ago, he realizes that what he sees is different from what he infers. The difference from the case of a brain looked at through a microscope is only one of degree: there is exactly the same need of inference, by means of the laws of physics, from the visual datum to its physical cause. And just as no one supposes that the nebula has any close resemblance to a luminous dot, so no one should suppose that the brain has any close resemblance to what the physiologist sees. (HK, 228, 229)

It is true that for many purposes we can forget the causal theory of perception and pretend that we have the power to behold physical objects directly. However, if we don't overcome this habit when we try to understand the relation of mind and brain, we go around and around in a circle of confusion regarding mental and physical. Our instinctive habit is to project the sensory data of our direct acquaintance onto the world outside our brains. That fosters the illusion that we have direct acquaintance with, and direct perception of, physical objects such as brains. We can't afford this illusion if we're to accomplish the business of this chapter.

> What I know without inference when I have the experience called "seeing the sun" is not the sun but a mental event in me. I am not immediately aware of tables and chairs, but only of certain effects that they have on me. The objects of perception which I take to be "external" to me, such as colored surfaces that I see, are only "external" in my private space, which ceases to exist when I die; indeed my private visual space ceases to exist whenever I am in the dark or shut my eyes. And they are not

"external" to "me", if "me" means the sum total of my mental events; on the contrary, they are among the mental events that constitute me. ... (HK, 225)

If we are to avoid going around in circles, we shall have to take science seriously. The notion that physical objects are made of matter was an assumption that took science a long way before running its course. We are now working under the hypothesis that physical objects are made of immaterial events. In the transition from one theory to the next, the role of sensory data in confirming theoretical conjecture is unchanged. We can't decide between a theory of matter versus a theory of immaterial events by direct inspection of the physical world. The theory of events is superior because it reduces the foundation physics to temporal succession as the sole hypothesis, while providing a role for sentient events and sensory data in the observation phase of experiments. To remain consistent, the hypothesis that a physical brain is a system of causally related quantum events cannot be mixed with assumptions that make the brain seem simpler, more familiar, or easier to perceive. This pertains likewise to tables and chairs. If we are unfamiliar with quantum events, then we are unfamiliar with tables and chairs, which are no more or less mysterious than black holes or what goes on inside the atom. What *is* familiar about tables and chairs, as Russell says above, is various effects they have on us. These familiar effects are among the sensory data that constitute our mental events.

In considering a chair, there is the physical chair made of quantum events on the one hand, and on the other hand, there are the familiar effects of the physical chair upon a person's experience. These effects make up what may be called "the phenomenal chair." On the physical side, there is no duplication of the chair into a "macro object" versus the chair as a system of "micro events." There is one consistent causal structure of quantum events, with no micro or macro about it. To harbor the notion that the physical chair includes some "overlay" of macro properties beyond the causal ordering of quantum events is to forfeit the consistency of the causal analysis. The physical chair has only two types of elemental constituents—quantum events, and the causal relations which order them. In that sense, the physical chair is extremely simple. Those elemental constituents establish layer upon layer of causal structure, according to the physics of nuclear, electromagnetic, and gravitational energies. In that sense, the physical chair is extremely complex. This pairing of simple and complex illustrates the limitless structural possibilities definable from a single relation. Our knowledge of the physical chair is confined to the theory of that chair as a causal structure of events. Turning to the phenomenal chair, our knowledge of it consists of our acquaintance with sensory effects presumably caused by the physical chair. Compared to the physical chair, the phenomenal chair has a richer variety of elemental constituents, including colors, the spatial relations belonging to our visual and tactile experiences, sounds, smells, and sensed time relations among these. With respect to number of elemental constituents therefore, the phenomenal chair is more complex than the physical chair. On the other hand, the structure of the phenomenal chair, which is due to phenomenal relations native to the sensory data, does not approach the structural complexity of the physical chair. Nevertheless, we piece together the theory of an extremely complex physical chair from the modest complexity of the structure of our sensory data. The extravagance of this inference is offset by the wide domain of the resulting theory, wherein the physics of virtually everything in the universe is exemplified in the physics of the chair.

Location of the human series in the brain sequence

> The standard conception of a brain is one of instantaneous extension in space, with no earlier-and-later involved in its composition. That is a brain without quanta. Such a brain has no place in our physics. Taking Special Relativity into account, the cortical surface is a set of contemporaries— "causal cousins"

· related only by their causal ancestry. Such contemporaries are also poised to beget common causal descendants. The location of the mind in the brain is resolved by tracing the causal lineage of human mental events to and from the *homuncular regions* of the cortex. These are the key causal locators of human mental events.

Brain scientists have mapped out a set of functional locations on the cortex called *projection areas*. These serve to pinpoint the location of the human series in the brain. The first two projection areas to consider are depicted by the motor homunculus and the sensory homunculus, which represent human-like forms that were first mapped out by Wilder Penfield. The topology of the human body is preserved in these forms, but geometric distortions in the drawing of the "little man" give him the appearance of a malformed fetus.

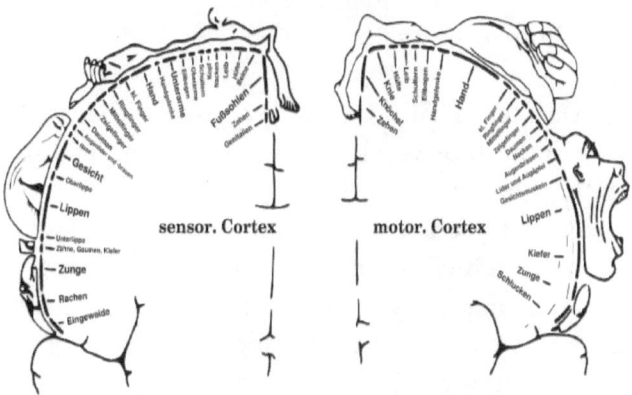

You can stimulate the motor homunculus with a probe and get the corresponding part of the body to twitch into action, like operating a puppet. You can stimulate the sensory homunculus to shortcut the more remote stimulus that is normally needed on the surface of the body to achieve the same sensation. Each moment of a human series of occasions has additional predecessors and successors that belong to the brain but not to the human series. Forking and convergence connect the human series to other cycles of the brain. Quanta that fork off from the human series to the motor homunculus provide control of bodily movement. Quanta from the sensory homunculus converge upon the human series, updating the body-image of somatic awareness. At cycles of ten-per-second, the sequence of cause-and-effect is as follows:

1. One human moment forks off via efferent quanta to many moments of the motor homunculus.

2. Effects are propagated along efferent nerve routes to the muscles.

3. Muscle action causes feedback signals along afferent nerve routes to the sensory homunculus.

4. Many moments of the sensory homunculus converge via afferent quanta upon the next moment of the human series.

During the tenth of a second between the two bounding moments of the above cycle, one human quantum also transpires, propagating the human series.

Other projection areas on the cortex have also been mapped out, which correspond to other sensory fields of human phenomenology. Patterns of excitation at the retina are reproduced at the visual projection area. Auditory experience also has a patch of cortical surface devoted to it. A mental event typically involves all the sensory modes at once. The distinct phenomenal sensory modes correspond to the distinct patches of cortex devoted to the organs of sight, sound and touch. As with the sensory homunculus, the visual and auditory projection areas are home to causal predecessors of the human series. From those cortical sites, the afferent system converges to a human percept, at which point the efferent phase of causal sequence is renewed.

Let us consider visual experience and its patch of cortex. In phenomenal vision, we have a spatially extended field of colored patches. The colorful visual field is part of a human mental event. As is the case with color, the inherent geometry of the visual field is given to the subject of experience. We can judge with remarkable precision the size and shape of colored patches given in our visual experience. A good example is the extraordinary precision by which we can judge a rectangle to have the pleasing height-width proportions of the *golden mean*. This is pure phenomenology. The ancients could judge with the same accuracy. Such judgment owes nothing to science. It is a judgment of ratio measurement that involves no scientific conceptions or assumptions. The visual projection area has a space-time metric based on *the second* as the standard unit of duration. Supposing the patch of visual cortex to be roughly circular and one inch in diameter, its space-like extent is approximately one-tenth of a nanosecond. The full spread of the subject's phenomenal visual field correlates to the full diameter of the cortical patch, so that half the visual field corresponds to half the cortical patch, or one-twentieth of a nanosecond. Proportionate size in the visual field is thus correlated to the metric unit of physics. This correlation is critical for an epistemological account of physical measurement, which requires sentient mental events in the laboratory, and sensory data that is phenomenally given to them. The correlation of phenomenal measure to the nanosecond span of this or that cortical projection area is reliant on psycho-physical experiments. Perceivable sensory fields are correlated to the unperceivable domain of physics. In the case of hearing, it is phenomenal *pitch* that correlates to the nanosecond span across the auditory cortex. We do not expand the domain of the perceivable by arriving at such correlations. Such results are obtained in the field of psychophysics, which correlates the qualitative data of subjective experience to the conjectural model of theoretical physics.

    The human series has direct access to vision, hearing, and tactile information at the cortical projection areas. Such *direct access* to information is unambiguous in our theory of physics. It means that select moments of the projection areas are immediate causal predecessors of a human moment. Each such predecessor connects to the human moment by a single quantum. Conversely, *direct action* by a human moment upon some moment in the region of the motor homunculus means that a single quantum connects the human moment to the homuncular moment. The homunculi are situated on the cortical surface as if to provide convenient test points for a technician to troubleshoot the sensory and motor systems. In normal operation, the cortical projection regions serve as staging areas for perception and control by the human series.

The stable brain is a propagation of synchronized time cycles, featuring a great range of frequencies and no doubt a great variety of cycle topologies. The stability of human experience and its dependence on the brain means that the human series must be embedded in supportive cycles of 10 Hz frequency. These cycles provide a base of causal routine for the human series, and they connect the human series at 10 Hz to the ladder of higher frequencies involved in brain function.

To conclude this chapter, we consider once more the laboratory setting of the psycho-physical experiment. We conceive the physical setting to be a system of causally ordered quantum events, accounting for the space of the laboratory and all the physical entities in it. In addition, we attribute phenomenological mental events to each of the people in the laboratory. The experiments indicate that the immediate causal predecessors of human mental events are physical events distributed across the cortical surfaces of human brains. This establishes the causal location of human mental events with respect to physical events. Thus, we rely upon prior knowledge of the purely physical configuration of the laboratory in order to discover the location of mental events. That prior knowledge, though routine and taken for granted, must itself be inferred from the sensory data presented to the sentient observers in the laboratory. That is, we rely on the phenomenal spatial relations of our own mental events to infer the causal location of physical events that make up the physical environment. The location of physical events is therefore defined in terms of causal order and known by inference, which is precisely the case for the location of mental events. Once physics is developed in sufficient detail to give a working knowledge of the brain, we are able to infer the causal location of our mental events with respect to physical events according to psycho-physical experiments. We must learn the location of physical events in some detail before we can learn the location of mental events, but that process of learning employs a single method and establishes physical location by a single criterion. The conjecture of causal order, in the service of predicting our sensory data, is the single means by which we know the physical location of either mental events or physical events.

Historically, it has always been assumed that a conscious mind and a physical body are each situated in time, but that a mind cannot be intelligibly conceived to occupy space as a physical body does. But space and its bodies must now be conceived as "strung out in time" according to the dictates of Special Relativity. The non-serial structure traditionally associated with causal order must be conferred upon time itself. Instantaneous spatial relations in physics, conceived according to the model of spatial relations in the visual field, are discarded. Spatially extended bodies are reduced to patterns of time-ordered events. This altered framework of science provides a meaning for the physical location of human mental events, and a basis for discovering that location by experiment. The overall effect is to make mental events seem more "physical." The distinction between mental and physical now hinges upon the assumption that mental events are characterized by sensory qualities, while physical events are not.

CHAPTER 7

Scientific Knowledge Characterized

Physical science constructs a causal model of the world for better predicting the patterns of qualities witnessed in human mental experience. The scientist has no privileged capacity to escape the confines of his mind to investigate the physical world directly. A predictive model is framed, tested, and refined solely based on phenomena witnessed in mental events. Scientific knowledge resides entirely in such models.

Some classical philosophers characterized matter as the unknown cause of sensation. Once we include spacetime (or a 4-D time manifold) among the theoretical constructs of physics, we can characterize the entire world beyond our sensory data as the unknown cause of sensation. We cannot start with incontrovertible evidence and proceed by logical deduction to crank out scientific knowledge. We need to supplement individual pieces of evidence with hypotheses and generalizations that surpass the evidence, in order to obtain theories that are useful for designing experiments and predicting their outcomes. Such generalities are drawn from a predisposition to believe that various observed regularities will continue to be observed.

> Broadly speaking, scientific method consists in inventing hypotheses which fit the data, which are as simple as is compatible with this requirement, and which make it possible to draw inferences subsequently confirmed by observation. ... (HK, 311)

The invented hypotheses to which Russell refers may be difficult to recognize as such. In hindsight, it is easier to judge that belief in *matter as a substance* was a hypothesis, since that belief has been discredited by science. The notion of a homogenous substance called "matter" is an example of the difficulty in recognizing the hypothetical status of beliefs we grow up with. Belief in matter has a strong grip on the imagination. Russell has said that people preserve their belief in matter as something "bumpable into." If I bump into a table in a dream, and remember that dream upon waking, I can grasp that the sensory experience of "bumping" does not in every case require an existing physical object. This helps me discriminate the sensory experience of bumping from "physical bumping." Furthermore, I can play with two sufficiently strong bar magnets and find that, with their poles oriented to repel, I cannot clap them together. The two magnets "bump" without making contact. This behavior of magnets can serve as a bridge to understanding how it is that science can possibly dispense with "hard matter". Ordinary objects are made of atoms. The nuclei of these atoms avoid contact with one another. They are shielded from one another by clouds of electrons. When two objects collide, their contact is better conceived as an electromagnetic bounce, like the bar magnets that bump without making contact. Turning to the sub-atomic particles, it is likewise intelligible that their mutual interactions can be described by nuclear forces without assuming that tiny material particles come into contact. Thus, a slight familiarity with magnets can go a long way toward an acceptance of the scientific demise of matter as a simple substance. Physicists are trained through common sense, and they remain realists in their attitudes regarding the physical world. They have not got rid of homogeneous matter without finding a suitable substitute. Russell and Whitehead have this same realism in their outlook on the physical world. Thus, in the analysis of the physical world into events and causal relations, those two types of entity are meant to bear the full load of commonsense belief in the existence of the physical world, a belief that is no longer well-served by the notion of a substance called "matter."

If you do manage to rid yourself of belief in material substance, you still face the mirage of empty space, which is an even greater obstacle to our understanding. Empty space is the receptacle for intangible magnetic fields, gravitational fields, and nuclear forces. It does not seem that you could relinquish belief in this static receptacle of space and still preserve a meaning for physical existence. Again, as in the case of matter, science does not divest itself of three-dimensional space without supplying a more adequate replacement. In this case, the concept of a space that exists "all at once" is abandoned in favor of a procession of intersecting time sequences, which accounts for a limiting velocity in the universe. Physical location becomes a matter of relative position in a causal sequence of events. The previous chapter explained how physical location is assigned to mental events in accord with this view of science. This chapter looks at the same situation from the standpoint of the mental events. Rather than inserting mental events into a given system of physical events, we consider mental events to be the known base of scientific observation, and the rest of the world a "scaffold" of cause-and-effect constructed around the mental events by systematic conjecture. From this perspective, mental events serve as the known origin for the causal location of physical events, and the contention of the previous chapter, that mental events have physical location, is a circumlocution and a foregone conclusion.

This does not detract from the validity of the previous chapter, but rather lends reinforcement to it. Whether we assume the validity of modern scientific theory and subsequently find a place for mental events within that theory, or whether we start with the sensory data of our mental events and explain scientific knowledge as an extrapolation from this data to a wider causal setting, we come full circle to a consistent analysis of scientific knowledge and the method of its acquisition.

Before science could come full circle, it had to start somewhere. It did *not* start by investigating how our sensory organs work, a project which could not get off the ground without considerable developments in the understanding of chemistry and electricity.

> Our perceptive apparatus, as studied by the physiologist, can to some extent be ignored by the physicist, because it can be treated as approximately constant. It is not, of course, really constant. By squinting I can see two suns, but I do not imagine that I have performed an astronomical miracle. (HK, 208)

An understanding of sense receptors, nerves, and brain could be postponed to a late stage of science since these organs function well without drawing attention to themselves. The early focus was on simple machines, such as levers, pulleys, and wedges, which give a mechanical advantage. Although the understanding of cause-and-effect would inevitably extend to the workings of human perception, it was first applied to the relations of external physical objects to one another. This focus on the mechanics of external bodies carried science to a sophisticated stage without shedding much light on human perception. Phenomenological data, such as colors, became marginalized in the scientific picture when it became apparent that they were superfluous to the mechanistic model. Unwittingly, this marginalized the sentient human being in the scientific view. With full confidence in the mechanistic model, investigating the role of the sentient human observer in science could even be put off indefinitely, since the observer is presumably just one more mechanism, of modest energy levels, on the backroads of physics.

A theory of physical mechanisms is fine and dandy, but if it finds no place for the sensory qualities we experience, then it offers the sentient observer no reason for believing it. Thus, we have suggested a perfectly good location in space-time for mental events in the preceding chapter, which fits the view of science as a theory of cause-and-effect. As Russell says, this is required if science is to be understood as an empirical study rather than an exercise in pure mathematics. From the vantage point of epistemology, which is concerned with *how* we know what we think we know, mental events of our sentient perceptual experience are the only events that *must* be included in a theory of space-time. Our mental events provide the proving ground for the conjecture of further events lying beyond human experience. For that reason, we deny that empirical science can omit mention of mental events, and explicitly adopt a view of science that incorporates mental events at the foundation of the theory. In order to integrate mental events into physical science, we can define "physical event" as "any event that is either a cause or an effect." This recognizes the fundamental role of cause-and-effect in the construction of scientific theory. Together with the understanding that mental events are the given effects for which a theory of causal explanation is constructed in the first place, we hereby induct mental events into the set of physical events.

> Percepts, considered causally, are between events in afferent nerves (stimulus) and events in efferent nerves (reaction); their location in causal chains is the same as that of certain events in the brain. Percepts as a source of knowledge of physical objects can only serve their purpose in so far as there are separable, more or less independent, causal chains in the physical world. This only happens approximately, and therefore the inference from percepts to physical objects cannot be precise. Science consists largely of devices for overcoming this initial lack of precision on the assumption that perception gives a first approximation to the truth. (HK, 209-210)

A "percept," as Russell uses the term, is the sort of mental event involved in perception of the external world. There are other sorts of mental events, such as in daydreaming, which do not involve an intent to perceive the physical environment. Percepts are of primary interest here as the mental events required for establishing knowledge of the wider physical world. In order to furnish scientific theory with the sensory evidence required for corroboration, we regard percepts as physical events crucial to the theory.

> If the percept is to be a source of knowledge of the object, it must be possible to infer the cause from the effect, or at least to infer some characteristics of the cause. In this backward inference from effect to cause, I shall for the present assume the laws of physics.

> If percepts are to allow inferences to objects, the physical world must contain more or less separable causal chains. I can see at the present moment various things—sheets of paper, books, trees, walls, and clouds. If the separateness of these things in my visual field is to correspond to a physical separateness, each of them must start its own causal chain, arriving at my eye without much interference from the others. The theory of light assures us that this is the case. Light waves emanating from a source will, in suitable circumstances, pursue their course practically unaffected by other light waves in the same region. (HK, 206)

I wish to make further use of the arrow diagrams as a graphic depiction of the scientific view of the world. Let us confine discussion to a single grand diagram which represents the entire physical universe, "the arrow diagram of the universe." This diagram is made up entirely of arrows which connect to one another. Every junction of arrows represents an individual event. Nothing happens in the universe but these individual events and the transitions that connect them. The arrows indicate the order in which the events happen. Some of the events in the diagram are the human mental events which comprise our phenomenological experience. Some of those events are percepts situated in the causal sequences described by the causal theory of perception.

A given mental event, without regard to its causal relations to other events, consists of a variety of sensory qualities organized into a unity of momentary experience. Some of the sensory qualities are relations, such as spatial relations among colored regions in the visual field. These relations provide the mental event with internal structure which affords analysis of the mental event into component parts. The arrow diagram represents the time relations of events dissected to the granular level implied by quantum theory. Since an individual mental event admits of logical analysis into whole and part, how can a mental event be represented in the arrow diagram as utterly simple and incapable of further analysis? The arrow represents the specific relation of temporal succession. The events located at the arrow junctions are irreducible with respect to *temporal* relations specifically, which means those events are not subject to further breakdown into parts that come before and after. The human mental event, as we have defined it, is reduced to a temporal moment, such that all component sights, sounds and feelings are simultaneous in one person's sentient experience, and no part comes before or after any other part For example, the spatial relations presented in the visual field are instantaneous and irreducibly spatial—we cannot reduce them to time relations as we have done for the space-like relations of physics. Since a mental event is a momentary composite of sensory qualities, it is appropriately represented as a temporally irreducible unit at the junction of arrows in the arrow diagram of physics.

Since we also identify the arrow as the *causal* successor relation, events connected by the relation are *causally primitive*. They are irresolvable by further cause-and-effect analysis. The proof is in the constructions of chapter 5. The limit of causal analysis has been reached. The final breakdown into physical units (moments and transitions) has been obtained for the coherent foundation of physics. Therefore, a mental event is individual in two ways. For itself it has the unity of sentient awareness. For physics it has causal individuality, as a unit of cause and as a unit of effect.

Our arrow diagram of the universe is meant to serve the imagination as a graphic depiction of the causal structure of our physical world. Russell contends that scientific knowledge is limited specifically to knowledge of causal structure, and secondly, that such knowledge rests upon conjecture. By contrast, our knowledge of sensory data, as pursued in phenomenology, is confined to that with which we have direct acquaintance. Phenomenology suspends judgment about the validity of any scheme of conjecture, causal or otherwise, that purports to provide a wider context for the realm of appearances. Hence, phenomenology does not set foot into the domain of science, and there is no overlap between phenomenological knowledge and scientific knowledge. There is, however, a crucial sense in which phenomenological knowledge, which Russell calls "knowledge by acquaintance," takes precedence over scientific knowledge. The descriptive use of language can only gain a foothold where we have acquaintance with recognizable entities that we can name. This is the case with phenomenological description, but it is not the case with physical description. We have no direct acquaintance with any scientifically defined entities. In general, we can only legitimately denote a scientific entity as "that which bears some conjectured causal relation to sensory data." If it is granted that science supplies the authoritative account of all physical bodies, then we do not have direct acquaintance with any physical body, contrary to our practical habits of thought and speech. According to science, knowledge of our physical surroundings is without exception mediated by neural signals. We distinguish a physical blow from an odor, or a flash of light from a roll of thunder, solely determined by which neural pathways are engaged in which firing frequencies. This uniform encoding of information is the source of all our opinions about the physical world. There is no bypassing this neural source of information—no alternate means of corroborating what our nerve signals tell us. One cannot adhere to science while rejecting this well-established doctrine of the human perceptual apparatus. Science began under the assumption that physical objects are directly perceived, and science has effectively disproved that assumption by establishing the role of neural transmission as the sole source of our perceptual information. If you hold a purely physicalistic doctrine of what exists, you are stuck in a paradox as to how perception happens at all, since a nerve cell is no more perceivable than any other physical entity. When this state of affairs is fully appreciated, one adopts a more conservative estimate of what can be known about the physical world. One is then amenable to the interpretation of science as a doctrine of causal structure inferred from the phenomenology of our mental percepts.

Here is the method that has governed the development of scientific knowledge:

1. Invent a hypothesis that predicts resultant events on the basis of control events.

2. Arrange for the control events to occur in a controlled experiment.

3. Observe whether the predicted results occur.

4. A successful result must be repeatable by independent teams of scientists.

In the first step, an initial hypothesis is required. This hypothesis must be formulated as a generalization which entails that certain events will follow upon certain others. The prediction might specify only a statistical outcome of many trials or cases. The merits of various hypotheses are judged by their predictive power.

Successful hypotheses developed through initially independent sciences have now been distilled (with the constructions of chapter five) into the hypothesis that the temporal/causal succession of events is all there is to physics.

Secondly, it is presumed that we can intervene in the causal order to ensure that a configuration of events under our control will occur. Occasionally we must wait for nature to bring about the conditions which will decide the truth of a prediction, as when an eclipse of the sun was used to confirm a prediction of General Relativity. More often we must actively bring about foreseeable events to serve as a controlled experiment, as when the first atomic bomb confirmed the conversion of mass to energy.

Thirdly, the outcome of an experiment must issue in an observation which decides the success or failure of the prediction. As Whitehead points out, every such observation is itself an event. In the terminology we are employing, the observational event is a mental percept situated in the wider purported causal order of events. In the case of an experiment in particle physics, the collisions induced by an accelerator leave artifacts on photographic film. After the film is developed, someone views the film and experiences visual percepts which are used to infer the outcome of the experiment. In this example, the elaborate chain of events between the particle collisions and the subsequent observation make it commonplace to say that elementary particles are not directly perceived but are inferred by means of theory. This wisdom generally evaporates when the observation of ordinary physical objects is considered, though scientific theory clearly implies that neural events and delays intrude between ordinary objects and our perception of them.

Finally, there is a demand for repeatability of experimental results. This requires of the physical world an intricate fabric of extremely dependable causal patterns. Causal patterns have been pursued to a level of detail that discloses irresolvable quantum events. There, where science runs out of pattern and predictability, it reaches a boundary to its knowledge.

The view of science proposed here, which is hopefully becoming clear, furnishes a new perspective to the contention of Chapter 2 that physics omits all sensory qualities from its theory. The scientific method equips physics to discover patterns of cause-and-effect that predict the sensory effects we experience. The results of this method have distilled the physical realm into units of occurrence which we are calling "events." If, as we should expect, physics adheres to its method, it can only deal with a mental event as a unit in a scheme of causal structure. The method provides no tool with which to delve into a causally simple event. As an example, consider the quality of redness. I can recognize what I call "redness," and I can verbally report its presence in my experience. The redness that I experience is logically simple—it has no analysis into parts. There does not seem to be any way of knowing whether the quality that I call "redness" is the same quality that you call "redness." Because redness is logically simple, neither of us can suggest any structural breakdown by which we might communicate a similarity or difference between the redness I see and the redness you see. Thus, for the purposes of science, which requires intersubjective agreement, a logically simple quality such as redness is completely useless. My verbal reporting of a red sensation, as a consistent response to consistent stimuli, is a different matter. Unlike the quality that I denote by it, the word itself, whether written or spoken, is a well-structured causal pattern, and as such, is fodder for scientific investigation.

Because our sensory data includes relations that bind other sensory items into structured wholes, we are acquainted with things that are *not* logically simple, such as the human visual field. This allows us to correlate the mathematical structure of a presented visual field with the conjectured structure of cause-and-effect in the physical world. This correlation is so intuitive that we initially assume that the visual field is a faithful representation of the physical environment. Science first modified this assumption by characterizing the physical environment as a spatial structure without sensory qualities. The representation of the colorless physical environment by the colorful visual field is therefore less faithful than was initially assumed. The next great modification to initial assumptions is the reduction of physical space to a structure of time-ordering relations. This means that the all-at-once spatial pattern of the visual field does not correlate to an all-at-once pattern of spatial relations in the physical environment. After these correctives from scientific theory are taken into account, what legitimately remains of the initial assumption that the visual field we experience is a faithful representation of our physical surroundings? Our intuitive projection of the visual field onto the physical environment is valid only to the extent that mathematical structure due to phenomenal relations of the visual field matches some mathematical structure due to causal relations among physical events in the environment. More generally, the structure disclosed in the full phenomenology of our senses furnishes the entire basis by which we articulate the scientific theory of causal structure.

The thesis of this chapter holds that science is restricted to conjectural knowledge of causal structure. The basic idea is that without conjecture, we are limited, as far as knowledge of the temporal world is concerned, to the phenomenology of sensory qualities. We depend upon conjecture for the rest. There is no shortage of conjectural material to work with. Before we engage in logical analysis or scientific discipline, we find ourselves firmly committed to various beliefs that go well beyond the sensory evidence. Santayana called such belief "animal faith." Today we are more likely to say that we are "hard-wired" for such belief. In any case, critical reflection exposes the conjectural nature of our dearest beliefs, including belief in other minds and belief in the existence of the physical world. In the light of this, we can abandon such belief and restrict ourselves to phenomenology and mathematics, or we can accept the non-demonstrable nature of our basic beliefs as final and inescapable. Opting for the latter, our aim is a consistent, coherent account of human minds in a physical environment. Science has already accomplished the difficult part. All that's needed is a judicious interpretation of current scientific knowledge.

In support of Russell's view that science gives us nothing but knowledge of causal structure, we can pursue a "positive" line of argument or a "negative" one. The positive line is constructive, building up current scientific theory from the single formal postulate of causal relations. The negative argument concerns the lack of any reasonable alternative. I will start with the negative.

I think it's fair to characterize the standard view of science by reference to the phrase "spatial-temporal causal framework," which Herbert Feigl has called the "the frame principle of science." The phrase begins with an homage to Special Relativity by its reference to space-time, which is commendable. We then come to the word "causal," which is crucial to Russell's point of view. Finally, we have the word "framework" to round out the phrase. Since no part of the phrase is inconsistent with the thesis of this book, what's the problem? The problem is redundancy, which though logically harmless, is symptomatic of the failure to make a conceptual breakthrough. We have the habit of conceiving the physical world as "space and what's in it." We are forced by scientific progress to modify that formula to "spacetime and what's in it." The concept of the physical world as an irreducibly spatial form of existence has a hold on the imagination, so that spacetime is conceived as a set of Newtonian-like spaces, one for each possible direction and speed. This is to rescue spatialized objects as the paradigm of physical existence. This is too complex, like the theory of epicycles, which tried to rescue the notion of a stationary earth by assigning highly peculiar orbits to the remaining planets. In place of "spatial-temporal causal framework" as the frame principle of science, I suggest "the causal order of events." What science requires from spacetime is its order, and the type of order it requires is causal order. Spacetime is not something over and above the causal order of events, and we should not let a redundant concatenation of terms obscure this. Furthermore, the term "framework," which is also redundant, connotes a physical world that is something substantial, or even material. This is exactly what is not needed. The world is a succession of discrete immaterial events. To summarize, the conventional view of science cannot dispense with the causal relations that suffice as the lone assumption in Russell's account. What distinguishes the conventional view is the determination to conceive even the most primitive physical entities as having geometric form, like the points of space or the particles of matter in Newton's physics. This compromises the straightforward interpretation of Special Relativity given in Chapter 5, which reduces space-time to a structure formed of purely temporal relations. There is a high cost involved in preserving geometric form for the fundamental entities of physics. Special Relativity must be reined in at small scales of distance, introducing difficulties at the boundaries of its domain. More importantly, there is the perpetuation of mystery regarding sensory qualities for the foundations of science and the mind-body problem. Thus, I contend that the standard view of science, as just described, is unnecessarily awkward and not likely valid. Greater rewards come from framing theoretical physics in accord with Russell's view that scientific knowledge can at best discover the "causal skeleton" of the world.

The constructive argument for the view that science describes only the causal order of events was initiated in Chapter 5, with the arrow diagrams provided to illustrate causal structure. I explained in the next chapter how physical location consists of relative position in the causal order, and in this chapter, how human perceptual events furnish the phenomenal evidence by which the causal order of any and all events is inferred. We shall now summarize and extend the interpretation of science in terms of causal order.

We hold fast to a standpoint that recognizes sensory qualities as the fabric of our direct experience. From a skeptical point of view, the sensory qualities are evidence of nothing but themselves. Phenomenology recognizes that limitation and explores its subject matter within that confinement. Accordingly, we recognize that conjecture is required to interpret sensory qualities as indicators of a physical world beyond the senses. In reviewing Newtonian science, we find that its material substance has been discarded, and only its structural characterizations survive as valid science. Special Relativity suggests that physical space itself is a structural pattern of the causal succession of events. Science employs various types of logical structure provided by mathematics to progressively "branch out" the structure of causality. At every stage of this scientific progress, sensory experience provides the predicted effects which alone can verify theory. Therefore, no matter what mathematical form a completed theory of physics might take, the mathematical expression of the theory can only designate a proposed causal structure of the world. The theory will imply a multiplicity of individuals related to one another by cause-and-effect. The individuals, which we have been calling "events," are only characterized by science as placeholders in a causal structure. Our mental percepts are among the events in any theory, since they are the only effects for which causes are sought in the first place. The scientific method gives no clue as to the intrinsic nature of events, and no clue as to the nature of the causal influence that one event can have upon another.

> I conclude that while mental events and their qualities can be known without inference, physical events are known only as regards their space-time structure. The qualities that compose such events are unknown—so completely unknown that we cannot say either that they are or that they are not different from the qualities that we know as belonging to mental events. (HK, 231)

An "ontology" is a logical description of what exists. The view of Russell and Whitehead has been called an "event ontology." With this label, the relations that connect events are left unmentioned. This is a symptom of the general reluctance to admit the irreducible role of relations in the logical analysis of anything. In Russell and Whitehead, the causal relation is as important as the event. In any case, we now wish to explore the adequacy of the event ontology for the formulation of scientific theory.

In the mathematical expression of physical theory, the world is reduced to number and order. To correlate the mathematics to the actual world, numbers must pertain to numbers of actual entities, and order must pertain to the ordering of actual entities. In our ontology, we have the causal ordering of discrete events as the basis of physical order, and we have the number of events, the number of causal transitions, and the frequency ratios formed, from which to derive further numbers that have physical significance. The general idea is to frame scientific theory as far as possible without positing further types of entity. Toward that end, we have constructed a 4-D time manifold and the common particles from the causal order of events, deriving the known mass-ratios of the particles from their graphs. We've found the fine structure constant in a diagram of 137 arrows. We've found the ratios inherent in time diagrams, together with constructions of the common particles, to suffice for the measure of space, time, energy, mass, charge. Even the "second" is an arbitrary unit of measure, since duration only has meaning in ratio comparison to other durations. It's reasonable to conclude that all the number and structure of theoretical physics is derivable from "the arrows of time." It can hardly be an accident that the simplest and most symmetrical patterns that time can make reveal the most pervasive entities and features of physics.

Some years ago, I came across the book *QED The Strange Theory of Light and Matter*, by the physicist Richard Feynman. In that book he explains, for a non-technical audience, quantum electro-dynamics. That is the theory of electromagnetism harmonized with quantum theory. Mr. Feynman was a primary contributor to that theory and to the theory of nuclear particle physics as well. Upon re-reading Mr. Feynman's book, I see that it inspired me to use arrow diagrams in this book. Mr. Feynman used diagrams to explain the quantum interaction of light and electrons, using distinct graphic symbols for light and electrons. He explained that such diagrams are not merely a concession to the modest technical background of his intended audience. Rather, diagrams of that type form the basis for mathematical calculations that constitute the rigorous exposition of the theory. Graduate students spend years learning mathematical "tricks" by which to comb general results from the welter of quantum actions represented by structural combinations of the two graphic symbols. Mr. Feynman explained that mathematics plays a crucial but auxiliary role. He expressed hope that the lay reader might understand the theory better than his graduate students! A true grasp of the theory comes from considering the very simple actions that make up the physical situation, while the intense use of mathematics applied to these actions is peripheral to a genuine understanding. Furthermore, Mr. Feynman finishes his book by giving his opinion that the rest of physics will follow suit with quantum electrodynamics in the general characteristics revealed by his diagrams. He suggested that the theory of electromagnetism is unified in one scheme of diagrams, while a similar scheme would depict the behavior of nuclear particles.

Because we have seen the electron constructed from the arrows of time, there is no need to use a separate graphic symbol for the electron, as in Feynman diagrams. Furthermore, a single system of arrows suffices for electronic structure and nuclear structure. Nuclear structure is more compact than electronic structure because it is of higher frequency. Also, a 6-D lattice arrangement is needed for the nucleus, while a 4-D lattice principle suffices for electron and neutrino formations. Thus, the arrow suffices as the only graphic element, it depicts a quantum, and everything in physics is built from quanta.

Our mental events are quantum events. This clashes with our preconception that quantum events are "tiny," either in size or causal efficacy. But since these tiny events comprise the whole universe, let us consider the significance of a single quantum event more carefully. Regarding spatial size, we can estimate the sphere of influence of a quantum event by multiplying its time period by the speed of light. For a period of one-tenth of a second we get about eighteen thousand miles—plenty of range for a human sentient event to gather concurrent influences from the surface of a cortex.

The causal efficacy of a quantum event owes more to its location in the causal order than to its individual energy assignment. The human organism consists of many cells and organs going about their provincial business of staying alive. In that sense, the human being is a society. Its structure includes the nervous system as the organ of control and feedback which governs the gross behavior of the organism. If there is any one causal position in the whole animal that is the seat of control, it is where information from all the senses is gathered and where a unitary influence on the voluntary muscle system can be exercised. This causal locus is the surface of the cortex, which is also the locus of brainwave activity.

Descartes made use of a piece of wax in order to bring his notion of physical existence into stark focus. I have a rock in front of me right now, which fits nicely in the palm of my hand. As a physical object, it will serve just as well as Descartes' piece of wax. My rock is made of quanta. The constituent quanta of the highest frequencies connect to form protons and neutrons, which combine to form nuclei. The nuclei combine with electron clouds to form complete atoms, which in turn combine to form molecules. The molecular patterns connect to form the rock. An arrow diagram of my rock would show how all its quanta are connected into a single elaborate sequence. The quanta are not undefined. Each quantum is an irresolvable step of time sequence. My rock is a propagating time sequence, made of temporal transitions connecting the moments to one another. As it is with the rock, so it is with my hand that holds the rock, my body, physical objects in general, and the universe at large.

The typical physicist today is under the innocent impression that he knows the essential nature of a rock in his hand from direct sensory perception without having made any conjecture at all. That innocent impression gives the physicalist a head-start in his pursuit of understanding the physical world. He is pre-equipped with the certainty that geometric shape and size are primary features of physical existence. But he is pre-equipped with the wrong topology, and his certainty is only psychological. He gets the wrong topology—spatial topology—from his own sensory data, which he cannot distinguish from the physical world, which he thinks he perceives. Thus, he is stuck with a spatial conception of the world. It will take the dramatic and incontrovertible collapse of spatial states and spatial configurations in the theory of physics to make him rethink his assumptions about what is perceivable and what is not.

At this point, Russell declares the mind-body problem to be solved. I will not quibble with that conclusion. Nevertheless, I will use one more chapter to take issue with Russell's non-committal attitude regarding the nature of events that are not human mental events.

Russell lamented that he could not communicate his solution to many people, especially philosophers. He attributed this to the ingrained habit of mistaking sensory data for a direct presentation of the physical world. An "imaginative leap" is required to overcome the habit. In my own case, several years of devoted study had produced no imaginative leap. When I came upon the two passages in *Human Knowledge* that I quoted near the end of Chapter 6, I was suddenly struck by the realization that my whole acquaintance with the physical world, including whatever sense I might have of my own body, is causally hemmed in by neural events to some region of my brain. With that, I suddenly let go of the spatially conceived world. In its place was a world of sheer temporal advance, of which my sentient experience formed a natural part. I've tried to place Russell's passages in a context that facilitates the same imaginative leap for the reader. In the final chapter, Whitehead may prove to be of further assistance.

Up until now the term "causal relation" has mainly been employed as a logical term needed to analyze causal structure, with events serving as the required relata. The final chapter will flesh out an interpretation of causal relations and events.

CHAPTER 8

The Solution

Science delivers only the bare causal pattern of events. Among these events are sentient occasions of human perception, which provide science with its observational data. When the remaining events required for the causal pattern are considered sentient occasions also, a coherent view of the world is obtained.

The final chapter belongs to Whitehead. Russell and Whitehead agree in the major structural feature of the solution to the mind-body problem, which is the analysis of the physical world into causal relations among physically featureless events, some of which are human phenomenological events. Experience lends itself to phenomenological description, while space-time does not. Hence, the sensory qualities of human experience are fundamental in the ontologies of both Russell and Whitehead. These qualities are implicated in contingent facts of the temporal world as ingredients of human mental events. Both Russell and Whitehead are devoted to belief in the existence of unperceivable theoretical entities of physics. They trim such scientific realism to belief in two kinds of entity—causal relations and events. The virtue of that specific form of conjecture is that it frames scientific knowledge with the fewest assumptions, and sensory experience assumes an appropriate role as the source of evidence by which theory is verified.

Russell maintains a strict agnosticism regarding the intrinsic nature of events that are not human mental events. At the same time, he makes it clear that sentient experience is the only sort of temporal existence we can hope to conceive. Supposing there to be some other kind of event than a sentient occasion of experience, it could only be conceived in the negative, as "not experience."

> I hold ... that the physical world is only known as regards certain abstract features of its space-time structure—features which, because of their abstractness, do not suffice to show whether the physical world is or is not different in intrinsic character from the world of mind. (HK, 224)

The unstated assumption underlying a typical discussion of "consciousness and the brain" is that the brain is the "known quantity," submitting nicely to scientific techniques, while consciousness remains a holdout against these techniques, presenting a mystery. From the viewpoint of this book, consciousness is the part of the brain we're familiar with, and the mysterious part is the rest of it. It is true that there is a great deal of well-established theory regarding the causal structure of brain events. That theory, along with the rest of science, amounts to an elaborate hypothesis regarding the causal order of events. What those events are, with the exception of our own mental events, is perfectly unknown, and the greater mystery falls upon those other events, rather than upon human consciousness. Consciousness is a complex instance of sentient experience. The complexity involves language and representation, volition and memory, and a causal order of brain events that is difficult to ascertain. While these complexities pose interesting problems, it is sentience in its bare simplicity that lends an air of mystery and misgivings to a physicalistic approach to the study of consciousness. As described in Chapter 2, sentience, or feeling, depends upon sensory qualities, and these qualities have been excluded from the scientific account since the advent of a purely geometric model of the physical world. In relation to that model, any sentience—even the simplest feeling—is a complete mystery. The geometric model turns out to be "all form and no content." The only source of content for scientific theory is the phenomenology of our sentient experience. Once that is fully appreciated, the geometric model is readily interpreted as the specification of a causal structure of events that improves the systematic prediction of our sensory percepts.

The outcome of our considerations is a very simple conception of the world. The fundamental temporal entities we have called "events." The class of events can be divided into our own mental events on the one hand, and other quantum events. We refer to the recognizable components of our own mental events as "sensory qualities." We have not offered any hypothesis about the components of other events, events which might be utterly simple and without any components at all. Last but not least, we require relations, which we have called "causal relations." These ordering relations form the causal structure of events, the ascertainment of which is the function of science.

As previously explained, if we choose to wait upon further scientific developments for information about the intrinsic nature of quantum events, we shall wait forever. For the purpose of forming a general conception of the physical world, we already have all the pertinent information we shall ever have. The options for a worldview are limited. We may regard other quantum events as occasions of experience analogous to our own; we may regard quantum events as an unknowable type of existence; or we may suspend judgment indefinitely.

The moment it struck me that I had been under the spell of what Russell calls "a presumed familiarity with the physical world," I automatically assumed that sentient events supply the content that is missing from the purely structural specifications of science. I was most likely preset for this response by my previous reading of Whitehead.

> For example, let the working hypothesis be that the ultimate realities are the events in their process of origination. Then each event, viewed in its separate individuality, is a passage between two ideal termini, namely, its components in their ideal disjunctive diversity passing into these same components in their concrete togetherness. There are two current doctrines as to this process. One is that of the external Creator, eliciting this final togetherness out of nothing. The other doctrine is that it is a metaphysical principle belonging to the nature of things, that there is nothing in the Universe other than instances of this passage and components of these instances. Let this latter doctrine be adopted. (AI, 235-236)

Whitehead picks up where Russell leaves off. He supposes that the basic events required by scientific theory are, like human occasions, unities of sentient feeling. Such a view fits the label "panpsychism," since it implies that mentality of a sort pervades all physical existence. In this case, the sort of mentality involved is sentience—the sort of temporal existence that involves component sensory qualities. Whitehead's view could thus be called "pansentience." Panpsychism in some form has probably been around forever. It is usually denigrated as superstition by champions of the scientific point of view. Panpsychism can foster a mode of too-easy explanation, such as "falling objects fall because they *want* to fall." This can distract us from the pursuit of better explanations, such as Newton's theory of gravity. Because of this history of obstruction to the progress of science, panpsychism arouses ridicule and rabid reactions. Nevertheless, Whitehead's view is rooted in modern science. Russell, who doesn't generally mince words, acknowledges Whitehead's view with courtesy and respect. Those who denigrate panpsychism in the name of science are captive to the outlook described in Chapter 2, and they have not come to grips with the interpretation of science established by Russell and Whitehead.

Whitehead devotes little effort to justify his point of departure from Russell's agnosticism regarding the intrinsic nature of events that are not human mental events. Before all the pieces fell into place for Whitehead, he had already railed against the belief in what he termed "vacuous actuality" —that the physical world is an insentient mechanism. He had been influenced by Henri Bergson, who protested the "spatialization of nature," and by F. H. Bradley, who also stressed the primacy of sentience in the composition of the physical world. Whitehead's intuition of panpsychism is evident in *Science and the Modern World*, but it took two more years to hit upon the interpretation of Special Relativity that transforms "spatialized nature" into a purely temporal succession of events. That development allowed Whitehead to translate his intuition of panpsychism into a logically rigorous conception of the world. He adduces the following principle in support of the sentient nature of all events:

> In framing a philosophic scheme, each metaphysical notion should be given the widest extension of which it seems capable. It is only in this way that the true adjustment of ideas can be explored. More important even than Occam's doctrine of parsimony—if it be not another aspect of the same—is this doctrine that the scope of a metaphysical principle should not be limited otherwise than by the necessity of its meaning. (AI, 237)

The above principle is invoked to settle the issue of the intrinsic nature of quantum events. These events are the ultimate realities turned up by scientific investigation. They form the basis for temporal existence. Sentient experience is one form of temporal existence—the only example we know by immediate acquaintance. To inhibit generalization from the known case to the unknown case is to complicate our understanding of the world for no reason. Therefore, it is prudent to adopt the simplest hypothesis available that suffices for the justification of our beliefs and the correlative interpretation of our sensory data.

It has been a standing argument against mind-like entities that they are *unlike* the spatial entities of physics, so that causal interaction between the two is unintelligible. But now 'the shoe is on the other foot.' A time series of human moments is well suited to instantiate the causal order, while the remaining moments of physical theory have no specified attributes whatsoever. It is these latter moments that now stand in need of causal compatibility with mind.

The arguments offered may seem slight in view of the enormity of the consequences. If the arguments hold, the physical world is an interplay of sentient experiences. If not, it is a throng of abstract theoretical objects. The latter conception has the force of habit in its favor, since it has dominated the scientific outlook for three hundred years. However, if Chapter 5 is correct, the meaning of Special Relativity has not yet worked its full transformation on the scientific outlook. The superstition of our time is that the causal order of events provides a refuge for the spatially conceived entities of Newton's time. Unless arbitrary limitations are imposed upon Special Relativity, any entity previously defined with spatial characteristics must now be defined as a pattern produced by a temporal succession of other entities. These latter entities are subject to the same redefinition, and the process ends with quantum events that have no spatial characteristics. We wind up with a universe of pure activity, which Whitehead calls "process." The notion of a physical entity as an instantaneous spatial configuration has no place in modern science. Once that is accepted, one can search in vain for some criterion bearing on the intrinsic nature of quantum events to weigh against the principle adduced by Whitehead in favor of nominating all events as "occasions of experience." This secures a strong argument for panpsychism. Further support comes when we adopt this result as a provisional hypothesis and employ it in the coherent interpretation of our sensory experience and our scientific beliefs.

At this point, we shall assume the sentient nature of all events, and proceed to the consequent interpretation of nature as a causal order of sentient events.

> Our consciousness of the self-identity pervading our life-thread of occasions, is nothing other than knowledge of a special strand of unity within the general unity of nature. It is a locus within the whole, marked out by its own peculiarities, but otherwise exhibiting the general principle which guides the constitution of the whole. This general principle is the object-to-subject structure of experience. It can be otherwise stated as the vector-structure of nature. Or otherwise, it can be conceived as the doctrine of the immanence of the past energizing in the present. (AI 187,188)

In this passage, Whitehead refers to one person's sentient experience as a "life-thread of occasions." An occasion is a sentient event, and a life-thread is a time-ordered chain of sentient events. One person's experience is thus an integral part of the causal order of events that constitutes nature. We have employed "causal relation" as a primitive term in the formulation of scientific theory. For a duration of human awareness, the causal relation orders the moments of a person's experience into a temporal stream of experience. The view that a person takes of cause-and-effect in this case bears on the possibility of meaningful human choice and action. At the same time, the character of cause-and-effect in connecting human occasions of experience will bear on physical causation in general, since all quantum events, according to our current assumption, are sentient occasions of experience analogous to our own.

Above, Whitehead offers three alternative wordings to describe the same general fact of causation as the relation that orders individual occasions to form a structured universe. These are: "the object-to-subject structure of experience," "the vector-structure of nature," and "the immanence of the past energizing in the present." The first of these appeals to phenomenology and the discernment of object-versus-subject within human experience. The second appeals to the directedness of time and the temporal propagation of events in scientific theory. The third refers to the influence of the past upon the present, which applies equally well to human subjective experience and the progression of physical events in general. By declaring the three principles equivalent, Whitehead shows that he is pursuing a general conception of cause-and-effect that underlies both the dynamics of human experience and the wider dynamics of the physical world.

Whitehead is forthright about his belief that human existence, as well as physical existence in general, is emotional and purposeful. We are not accustomed to hearing such talk from a scholar of modern science. We are inculcated with the idea that scientific investigation discloses nature as a mix of random and automatic actions that do not admit of explanation in terms of purpose. Of course, if "sentience" is written out of the scientific lexicon, as it was in the Newtonian framework of matter-in-motion, purpose finds no place in scientific dialog. If, on the other hand, sentient occasions of experience form the basis of physical existence, it is no longer necessary to dismiss out-of-hand the discussion of purpose in a scientific context.

Whitehead says that physical science is an abstraction derived from more concrete fact. Mr. Gustav Bergmann has suggested that the word "abstract" should be avoided altogether in the context of logical analysis, and I have for the most part taken his suggestion to heart. The word has become an "escape valve" that is used to avoid coming to terms with relations and logical structure. We can restate the proposition that "physical science is an abstraction derived from more concrete fact" as follows: "The mathematical expression of physical theory refers to the logical structure of physical events as ordered by temporal-causal relations." If Richard Feynman is correct, the abstractions of physical theory can be specified in the form of diagrams, which depict nothing but logical structure. To construe an arrow diagram as a representation of the actual world, we require a correspondence between the junction points and arrows of a diagram with individuals and relations thought to comprise "concrete fact." This has led us to conceive sentient events and the manner of their effect upon one another as the elemental components of concrete fact. For the moment, let us suspend interest in causal connections to focus on the individuality of a sentient occasion.

> *Individuality.* The individual immediacy of an occasion is the final unity of subjective form, which is the occasion as an absolute reality. This immediacy is its moment of sheer individuality, bounded on either side by essential relativity. The occasion arises from relevant objects and perishes into the status of an object for other occasions. But it enjoys its decisive moment of absolute self-attainment when it stands out as for itself alone, with its own affective self-enjoyment. The term 'monad' also expresses this essential unity at the decisive moment, which stands between its birth and its perishing. The creativity of the world is the throbbing emotion of the past hurling itself into a new transcendent fact. It is the flying dart, of which Lucretius speaks, hurled beyond the bounds of the world. (AI, 177)

For Whitehead, the individuality of an event is an individuality of subjectivity and feeling. This is also how common sense conceives the basic individuality of one person's mind. For physics on the other hand, the individual event is defined entirely by its relations to other events, since events only serve physics as the bare individuals required for causal structure. The individuals of physics are "bare" because the scientific method offers no clue as to their interpretation. Whitehead proposes an interpretation of physical individuals that is most similar in the philosophical tradition to Leibniz' *Monadology*. Monads, as the elemental entities, have the individuality of subjective feeling commonly ascribed to a human mind. For both Leibnitz and Whitehead, this subjectivity is the mode of individuation by which the temporal world is split into a multiplicity of entities.

This way of conceiving individuation supplants spatial extension as the basis for analyzing the world into whole-and-part. It is then possible for Leibniz and Whitehead, each in his own way, to explain without circular reasoning, physical space as a system of relationships among sentient individuals. For Leibniz, this system of relationships is a correspondence between monads with respect to the variations in their constituent sensory qualities. This correspondence is implemented by God when he created the monads at the beginning of time. For Whitehead, the constituent properties of a monad only become determinate as a response to the causal influences of other freshly completed monads constituting the immediate past. Creation is thus distributed among the individual acts of individual moments, rather than consolidated in one act which determines the entire course of events. Whitehead is accounting for the causal independence of contemporary events implied by Special Relativity and Quantum Theory. Leibniz was adhering to the scientific determinism of his day—a determinism which *did* imply a pre-established course of events and *did* restrict the efficacy of purpose to a one-time act of creation. Given the constraints of scientific determinism, it is uncanny that Leibniz conceived a monadology so much like Whitehead's. They are brought into congruence when the causal agency that is attributed by Leibniz to a single divine act of creation is disseminated among events as the causal agents of their own immediate progeny. This casts the universe as a process of stepwise, piecemeal creation. Each transition is an instance of causal relation connecting event to event, and the causal order of nature is the pattern laid down by these transitions.

> The actualities of the Universe are processes of experience, each process an individual fact. The whole Universe is the advancing assemblage of these processes. ...

> ... Any set of occasions, conceived as thus combined into a unity, will be termed a nexus. ... When the unity of the nexus

> is of dominating importance, nexus of different types emerge, which may be respectively termed Regions, Societies, Persons, Enduring Objects, Corporal Substances, Living Organisms, Events, with other analogous terms for the various shades of complexity of which Nature is capable. ...
>
> The causal independence of contemporary occasions is the ground for the freedom within the Universe. The novelties which face the contemporary world are solved in isolation by the contemporary occasions. There is complete contemporary freedom. It is not true that whatever happens is immediately a condition laid upon everything else. Such a conception of complete mutual determination is an exaggeration of the community of the Universe. The notions of 'sporadic occurrences' and of 'mutual irrelevance' have a real application to the nature of things. (AI 197, 198)

Regarding the various shades of complexity of which Nature is capable, I wish to emphasize that such complexity is uniformly analyzable as variations of structure arising from sequential causal relations between events. For Whitehead, such events are, without exception, "occasions of experience," or just "occasions." An occasion is momentary, arising and perishing in its fixed location in the causal order. An occasion is a moment that neither endures nor recurs. As we consider various structural formations with reference to their common names, such as "persons," "societies," and "living organisms," we should bear in mind the uniform analysis of these assemblages into discrete units of experience. Otherwise, the analysis "goes soft."

We are using a single moment of one person's sentient awareness as the prototype for all events. It serves as the source of meaning for "the intrinsic nature of quantum events." One person's sentient experience over time, considered in isolation from other events comprising that person's brain and body, is commonly called a "stream of consciousness." It is, according to many religions, what survives the death of the body to host the pleasures or pains of an afterlife. It is a domain of feeling, variegated by qualitative features that furnish the subject matter of phenomenological description. Whether the feeling is rudimentary, as when awareness dwindles to a fog with the onset of an anesthetic, or whether the feeling is complex and sophisticated, as when a person considers the meaning of a proposition expressed in language, the ingredient objects of awareness are what we have referred to as "sensory qualities."

It is difficult, if not impossible, to isolate, via introspection, a "present moment" of experience. The "now" is like a moving target. On the other hand, it seems we cannot escape the present moment to target anything else. Earlier, when we "quantized" personal experience into discrete "drops" of roughly one-tenth of a second duration, we relied on psycho-physical experiment, which made the analysis dependent upon the conjectural basis of science. Furthermore, we adduced the phenomenological experience of judging earlier-versus-later as a basis for equating personal experience with a time-series of momentary experiences. That pleasing result has an air of construction and inference about it. It appears that through introspection alone personal experience cannot be sorted into component time relations versus momentary events. Even if that could be done, the time order established by that method could only account for private threads of experience causally isolated from one another. In that case it is advisable to regard "the causal relation" as strictly hypothetical, even as applied to the connectivity of one person's experiences over time. Recall Whitehead's doctrine that "each metaphysical notion should be given the widest extension of which it seems capable." In this case, the metaphysical notion is *the causal relation*, which represents the whole conjectural basis of scientific knowledge in distilled form. If we forego any special exemption regarding the conjectural status of the causal relation, even when it is invoked as the ordering principle of one person's stream of experience, then we place knowledge of one's personal continuity over time on the same footing as scientific knowledge. As Russell was quoted earlier, the scientific determination of causal order is based upon the phenomenological judgment of earlier-versus-later, but only as a first approximation to the truth. The initial judgment may be amended by scientific theory and measurement. The subjective judgment of time order—even the time order of one's private train of thoughts—is tainted with inference and subject to error. We do not know any instance of the causal relation the way we know a sensory quality. That takes proper account of the historical discussion of sensory qualities that led to the skeptical conclusions of David Hume. Hume found that he could logically deny any interpretation of sensory data that served to establish either the self, other persons, or a physical world. This skepticism reduced sensory experience to "solipsism of the present moment." The lesson to be taken from Hume is that any practical interpretation of sensory data relies upon one or more tenets of belief that are not logically entailed by the available evidence. The fact that routine interpretation of our sensory data is crucial to the meaning of our lives does nothing to change this. It is essential that the subjective privacy of feeling, which characterizes the individuality of one occasion, be connected to other occasions by a relation that does not compromise the principle of individuation. The logical distinction between individual occasions and the causal relation which orders them is what obliges us to *posit* the causal relation.

> *The Human Body*. But this analogy of physical nature to human experience is limited by the fact of the linear seriality of human occasions within any one personality and of the many-dimensional seriality of the occasions in physical Space-Time.

> In order to prove that this discrepancy is only superficial, it now remains for discussion whether the human experience of direct inheritance provides any analogy to this many-dimensional character of space. If human occasions of experience essentially inherit in one-dimensional personal order, there is a gap between human occasions and the physical occasions of nature.

> The peculiar status of the human body at once presents itself as negating this notion of strict personal order for human inheritance. Our dominant inheritance from our immediately past occasion is broken into by innumerable inheritances through other avenues. Sensitive nerves, the functioning of our viscera, disturbances in the composition of our blood, break in upon the dominant line of inheritance. In this way, emotions, hopes, fears, inhibitions, sense-perceptions arise, which physiologists confidently ascribe to the bodily functionings. So intimately obvious is this bodily inheritance that common speech does not discriminate the human body from the human person. Soul and body are fused together. ... (AI, 189)

The above passage is dealing with the fact that a person's stream of experience is not a causally autonomous series of occasions but is causally engaged with a confluence of non-human occasions at the surface of the cortex. These latter occasions contribute to the structure of the human organism, but do not belong to the human series of occasions that constitutes our inimitable experience. Each member of a human series has direct causal influence upon its successor, forming the dominant line of inheritance. Without further causal influences impinging on the members of a human series, the series would be autonomous, and the human sentient mind would be causally disengaged from the body. That is not compatible with an account of scientific knowledge that relies on human percepts for its empirical confirmation.

I used the following quote previously in connection with the phenomenology of time. I think it's worth repeating in the current context with a view to making sense of causal relations.

> Non-Sensuous Perception. ... Gaze at a patch of red. In itself as an object, and apart from other factors of concern, this patch of red, as the mere object of that present act of perception, is silent as to the past or the future. How it originates, how it will vanish, whether indeed there was a past, and whether there will be a future, are not disclosed by its own nature. No material for the interpretation of sensa is provided by the sensa themselves, as they stand starkly,

barely, present and immediate. We do interpret them, but no thanks for the feat is due to them. ...

> In human experience, the most compelling example of non-sensuous perception is our knowledge of our own immediate past. I am not referring to our memories of a day past, or of an hour past, or of a minute past. Such memories are blurred and confused by the intervening occasions of our personal existence. But our immediate past is constituted by that occasion, or by that group of fused occasions, which enters into experience devoid of any perceptible medium intervening between it and the present immediate fact. Roughly speaking, it is that portion of our past lying between a tenth of a second and half a second ago. It is gone, and yet it is here. It is our indubitable self, the foundation of our present existence. Yet the present occasion while claiming self-identity, while sharing the very nature of the bygone occasion in all its living activities, nevertheless is engaged in modifying it, in adjusting it to *other* influences, in completing it with *other* values, in deflecting it to *other* purposes. The present moment is constituted by the influx of *the other* into that self-identity which is the continued life of the immediate past within the immediacy of the present. (AI, 180, 181)

The first paragraph could have come from the writings of David Hume, summarizing as it does the requirement for interpretation in order to break out of a solipsism of the present moment. The second paragraph expands the account with a healthy dose of such interpretation. Most people will have no trouble with Whitehead's interpretation of the immediate past as an influence upon the present, since it is a straightforward description of the purposeful manner in which we all conduct our daily lives. The description would even be trivial if it did not contradict the prevalent doctrine of scientific explanation that excludes purpose as a factor in determining the course of physical events. This doctrine was consolidated with Newton's deterministic scheme of matter in motion. As that scheme has given way to the indeterminism of quantum events, the burden of scientific explanation has shifted to the workings of random chance. The exclusion of sensory qualities from the physical scheme has remained intact, in which case it makes no sense to consider "purpose" as playing a role in the insentient activities of nature. On this view of nature, we can either interpret human action as a product of random chance, or we can attempt to construe purposeful human action as an exception to scientific explanation. On the other hand, Whitehead supposes that every action of nature arises from, and issues in, sentient experience. On this view, the efficacy of purpose that we claim for our own personal actions is, without reason to think otherwise, fundamental to all causal action. A person is just one causal thread of sentient occasions among others that comprise nature, and a person's activity is representative, in fundamental respects, of any causal process.

The order of exposition that I have chosen began by setting forth a problem in terms of the dissociation of sensory qualities from the scientific outlook. This was then remedied by construing science as the ascertainment of a causal order of events that best predicts our sensory percepts. With Russell at the helm, the primary emphasis was on phenomenological percepts as *effects*, with the external world providing the *causes*. That much is certain if scientific theory is to be susceptible of empirical verification. Once we have established a coherent role for sensory qualities in the physical world, we can then attribute sentience and feeling to the physical world without scientific misgivings. This in turn makes it reasonable to suppose that efficacy of purpose is a causal factor in determining the course of physical events. Since science teaches that an individual quantum event is not entirely the resultant of external causal factors, and since we are supposing that any such event is a sentient occasion of experience, it is reasonable to speculate that each sentient occasion includes a component of self-determination based on purposive feeling. In that case, each occasion is not just a sentient entity, but also a sentient *entelechy*, engaged in its own formation and fulfillment. This is Whitehead's view.

> *Objects.* —The process of experiencing is constituted by the reception of entities, whose being is antecedent to that process, into the complex fact which is that process itself. These antecedent entities, thus received as factors into the process of experiencing, are termed 'objects' for that experiential occasion. Thus primarily the term 'object' expresses the relation of the entity, thus denoted, to one or more occasions of experiencing. Two conditions must be fulfilled in order that an entity may function as an object in a process of experiencing: (1) the entity must be *antecedent*, and (2) the entity must be experienced in virtue of its antecedence; it must be *given*. Thus an object must be a thing received, and must not be either a *mode* of reception or a thing *generated* in that occasion. Thus the process of experiencing is constituted by the reception of objects into the unity of that complex occasion which is the process itself. The process creates itself, but it does not create the objects which it receives as factors in its own nature.
>
> 'Objects' for an occasion can also be termed the 'data' for that occasion. The choice of terms entirely depends on the metaphor which you prefer. One word carries the literal meaning of 'lying in the way of', and the other word carries the literal meaning of 'being given to'. But both words suffer from the defect of suggesting that an occasion of experiencing arises out of a passive situation which is a mere welter of many data.
>
> *Creativity.* —The exact contrary is the case. The initial situation includes a factor of activity which is the reason for the origin of that occasion of experience. This factor of activity is what I have called 'Creativity'. The initial situation with its creativity can be termed the initial phase of the new occasion. It can equally well be termed the 'actual world' relative to that occasion. It has a certain unity of its own, expressive of its capacity for providing the objects requisite for a new occasion, and also expressive of its conjoint activity whereby it is essentially the primary phase of a new occasion. It can thus be termed a 'real potentiality'.

> The 'potentiality' refers to the passive capacity, the term 'real' refers to the creative activity, where the Platonic definition of 'real' in the *Sophist* is referred to. This basic situation, this actual world, this primary phase, this real potentiality—however you characterize it—as a whole is active with its inherent creativity, but in its details it provides the passive objects which derive their activity from the creativity of the whole. The creativity is the actualization of potentiality, and the process of actualization is an occasion of experiencing. Thus viewed in abstraction objects are passive, but viewed in conjunction they carry the creativity which drives the world. The process of creation is the form of unity of the Universe. (AI 178, 179)

We are exploring the idea that the universe is a causal process individuated into discrete temporal transitions from one or more sentient occasions to one or more other sentient occasions that are next in causal order. A complete arrow diagram of the universe would represent the detailed structure of spacetime as a concatenation of discrete steps of temporal transition. We wish to better understand the causal relations represented by the individual arrows themselves. To that end, we may focus our attention on the rudimentary case of causal structure depicted in the following diagram.

Causal X

The diagram helps to crystallize an overview of the component ideas of this book. Each arrow stands for a discrete causal/temporal transition from one sentient occasion of experience to another. Quantum theory dictates that the transitions be discrete.

Special Relativity requires the joining and splitting of separable time sequences. Without additional arrows to connect this diagram into a wider context of space-time, the five occasions involved in the diagram might equally well be part of a star, a piece of dust, or a human brain. The diagram depicts an ample fragment of space-time to illustrate Whitehead's remarks about "Objects" and "Creativity."

The occasion at the center of the diagram, which we may call "the central occasion," is, from a strictly physicalistic point of view, a typical uninterpreted individual at the base of all existence—a causally primitive event. Postulation of such individuals is adequate for the construction of a predictive model of cause-and-effect sequences. By relying on uninterpreted individuals, the physicalistic scheme provides primitive entities that are unknown and unimaginable. But we have found that our own mental events serve very appropriately as individuals of this causal structure, with mental percepts providing the theory of physics with testimony of empirical verification. We then found reason to believe that all the individuals are *occasions of feeling*, generalizing from our own mental events. This provides physics with an intelligible interpretation. Physicalism is a valid approach to scientific discipline in regard to the purely structural import of scientific claims. It oversteps with two unwarranted assumptions: firstly, that "physical" means something more than "cause-and-effect"; and secondly, that causes and effects must be something other than mental events.

If we consider the above diagram in isolation, it serves as a touchstone to the elemental facts of time and causation. An arrow represents the before-and-after pairing relation. The central occasion represents a "now" of momentary experience. The temporal "now" belongs to each occasion in its essential individuality. (The "now" is not a relation that defines a "same moment of time" for two or more occasions. There is no such relation according to Special Relativity, and no two quantum events happen at the same time. That is the "breakdown of simultaneity.")

The two arrows leading to the central occasion represent the combined influence of the past upon the determination of the central occasion. The entire causal past of this central occasion is mediated, as a formative influence, by just two occasions. Whatever "data" those two occasions provide to the experience of their mutual successor exhausts what Whitehead calls "the immanence of the past" in the central occasion. What is the nature of the data provided? The only sort of data we know is the qualitative phenomena ingredient in our own experience. Hence, we surmise that the phenomenological constitution of completed occasions becomes available as data to the arising experience of their immediate successors. This constitutes a breach in the privacy of individual occasions. This breach is depicted by an arrow, and we now have a conception of primitive causal relations wherein the phenomenological constituency of each completed occasion is rendered for appropriation by its immediate successors.

Any causal arrow depicts a case of "direct immanence." One may then define a transitive version of immanence such that any occasion is not only directly immanent in its immediate successors but generally immanent in the successors of its successors, and so on. Space-like relations are then defined for two contemporary occasions by reference to other occasions that are immanent in both of them—that is, by reference to other occasions sufficiently ancestral as to be in the common past of both contemporaries. Thus, the theory of spatial order is reduced to the structural consequences of immanence.

> In the formation of each occasion of actuality the swing over from re-enaction to anticipation is due to the intervening touch of mentality. Whether the ideas thus introduced by the novel conceptual prehensions be old or new, they have this decisive result, that the occasion arises as an effect facing its past and ends as a cause facing its future. In between there lies the teleology of the Universe.

If the mental activity involves no introduction of ideal novelty, the data of the conceptual feelings are merely eternal objects already illustrated in the initial phase of re-enaction. In that case, the re-integration with the primary phase merely converts the initial conformal reception into the anticipation of preservation of types of order and of patterns of feeling already dominant in the inheritance. There is a reign of acquiescence. In this way, a region of such occasions assumes the aspect of passive submission to imposed laws of nature. But when there is conceptual novelty made effective by its re-iteration and by the added emphasis on it throughout a chain of coordinated occasions, we have the aspect of an enduring person with a sustained purpose originated by that person and made effective in that person's environment. Thus in this case the anticipation of kinship with the future assumes the form of purpose to transform concept into fact. In either case, whether or no there be conceptual novelty, the subjective forms of the conceptual prehensions constitute the drive of the Universe, whereby each occasion precipitates itself into the future. (AI 193, 194)

In respect to the above passage, Whitehead uses the terms "prehensions" and "eternal objects." A "prehension" is a relation by which an occasion-in-process appropriates phenomenological data that will characterize its completed constitution. It is a grab for ingredients by the creative process in forming an individual occasion. The discernible ingredients of experience are what we have been calling "sensory qualities" or "phenomenological data." Whitehead calls them "eternal objects," alluding to Plato's conception of a timeless realm of possibilities implicated in the contingent facts of the temporal world. Individual occasions serve in our ontology as irreducible "particulars." The subjectivity of these particulars requires correlative objects. All occasions are alike in their basic subjectivity, so whatever intrinsic differences there might be between two occasions owes to a discrepancy in their respective objects. The sort of objects that account for qualitative similarities and dissimilarities are commonly called "universals." Each completed occasion is a fact consisting of a "particular" qualified by "universals," and more specifically, a subjectivity qualified by phenomenological data. Unless time is thought to create phenomenological data out of nothing, it is qualitative data, as potential for experience, that is prerequisite for time. The view that temporal process establishes contingent facts presupposes a wider range of unactualized possibilities. Sensory qualities participate in the temporal process as objects for actual occasions of experience. As a "universal," a quality is instantiated in some occasion, which gives that quality a definite location in spacetime. Just as red and green are mutually exclusive in the coloration of some patch of your visual field, any instance of a quality excludes various alternatives in the opportunity of that moment. A quality that is never experienced has no location in time, and the same holds true for any structured complex of qualities that is never experienced. Hence, unactualized possibilities have no location in time. That explains the term "eternal objects." The domain of eternal objects involves facts about the differentiation of objects among themselves and facts of structure due to relations that are native to the domain of objects. The color solid discussed in Chapter 1 is a good example. Several facts about the color solid were discussed—for example, that yellow is "fenced off" from blue in the hue circle by red and green. Such a fact pertains strictly to relations between colors and involves no essential reference to time. The color solid in its entirety is a complex eternal object. Enough is known about the finite complexity of human phenomenological experience to conclude that a person cannot experience all the colors of the color solid at one time. Since time is a succession of sentient occasions, and a phenomenal object has location in time by being ingredient in some occasion, then unless there are color-sensing occasions that outperform human beings, the color solid as a whole is an object without location in time. Yet we can piece together the facts about the color solid, and graphic artists make routine use of its organization in the process of selecting colors. This example is intended to remove any mystery or confusion about Whitehead's term "eternal object." Human understanding is not exclusively devoted to ascertaining contingent facts of time. Another example is pure mathematics, in which the vicissitudes of time are again beside the point. Mathematics delineates possibilities of structure, or what one might call "facts of form." The applicability of mathematics to the actual world, while crucial to science, is incidental to pure mathematics. Here we have a body of knowledge that has grown so large as to be beyond the ken of any one person. It has even been proved infinite in a way that cannot be systematically completed. Therefore, the logically conjoined object of mathematical investigation has no location in time. It is in this sense that the object of mathematical inquiry, like the color solid, can be considered "out of time," "timeless," or "eternal."

Now we resume the analysis of temporal process in terms of the prehension of eternal objects by sentient occasions. The temporal process is quantized into steps, just as sentient experience is quantized into occasions. We see a duplication in terminology at this point, such that "temporal process" and "sentient experience" have come to mean the same thing. The common referent is the actual world, analyzable into momentary, transitory individuals. The qualitative nature of the actual world is provided by the diversity of eternal objects—the potential objects of experience. If an object is appropriated by an occasion from one of its immediate causal predecessors, this appropriation is termed a "physical prehension." Physical prehensions are depicted by our causal arrows, the primitive relations responsible for causal order. If the data is a temporal novelty, prehended simply by virtue of its being a possibility, the appropriation is termed a "conceptual prehension." The latter is named after the conceptual facility of human mental experience, which seems to range more freely among the possible objects of thought than previous experience alone would explain. The generalized hypothesis is that an occasion in process of formation is conditioned first and foremost by its past via physical prehensions of its immediate predecessors, and optionally, by a kinship of the immediate past to related untried possibilities via conceptual prehensions.

The provisional definition of prehensions may be expanded by proposing positive and negative prehensions, which include or exclude an eternal object, respectively. Consider the arising moment of the central occasion reacting to its two immediate predecessor occasions. It is engaged in modifying them, in adjusting them to other influences, in completing them with other values, in deflecting them to other purposes, as Whitehead says. This involves selective inclusion and exclusion of the objects inherent in the causal parents. If the same distinction between positive and negative is applied to conceptual prehensions, then all possibilities are prehended by each occasion, either positively or negatively, since the inclusion of some involves the exclusion of the rest. With this expanded formulation, selection by an occasion from unlimited conceptual possibilities becomes more intelligible. Previous selections, prehended from the past, are connected to the entire realm of eternal objects by relations native to that realm, relations of similarity and contrast, for example. These native relations order the realm of possibilities according to what is more closely related, versus what is less closely related, to the accessible constituents of the causal parent occasions. This provides a principle of limitation that narrows selection from the overwhelming variety of all possibilities. It also seems to describe the workings of human imagination. A suitable metaphor suggests itself, that the temporal world has a foothold in the limitless territory of possibility, and that the creative process of which Whitehead speaks is engaged in progressive exploration of that territory.

If we now shift the focus of our attention away from the central occasion of the diagram, and outward to its wider causal context, we envisage the structural possibilities afforded by iterated causal relations, which returns us to the theory of spacetime and physical science. The structural possibilities are only broadly restricted by the rules of arrow concatenation—rules which only codify the notion that time is an accretion of irrevocable events.

Chapter 5 shows that a unified physical theory can be based on "the arrow diagram of the universe." Symmetry was the only organizing principle that governed the initial constructions. Thus, symmetry has assumed the role of "force" in physics, explaining the regularities of "law." But it is doubtful that the variety of biological lifeforms comes from symmetry alone. Why should select patterns of temporal succession prevail and not others? Why should there be any events or causal patterns at all? Such questions arise from the frustrations and satisfactions of sentient beings, so it is no mystery that a mathematical representation of causal structure cannot give the answers. If an event is conceived to be an insentient occurrence or held to the status of an uninterpreted mathematical construct, there can be no intrinsic reason for it, and it is absurd to ask why it occurred. The same holds true for cause-and-effect relations and the patterns built up from them. An interpretation of scientific knowledge that incorporates sentient beings is required for a concept of nature that supports *reasons for things*. We are presently engaged in such an interpretation, wherein nature is a community of sentient occasions—the "concrete reality" of which the causal structure is specified by diagrams. This fuller conception of the physical world provides "reasons why." To begin with, each individual event, being an instance of subjectivity and feeling, is something *for itself*—something for its own sake. Its measure of satisfaction constitutes, in the most ordinary sense, a reason for its existence. Beyond that, we might ask why events form peculiar patterns of causal succession. Whitehead provides an answer in terms of "societies."

> We now pass on to the general notion of a *Society*. ...

> A Society is a nexus which 'illustrates' or 'shares in', some type of 'Social Order'. 'Social Order' can be defined as follows: -- 'A nexus enjoys "social order" when (i) there is a common element of form illustrated in the definiteness of each of its included actual entities, and (ii) this common element of form

arises in each member of the nexus by reason of the conditions imposed upon it by its prehensions of some other members of the nexus, and (iii) these prehensions impose that condition of reproduction by reason of their inclusion of positive feelings involving that common form. Such a nexus is called a "society", and the common form is the "defining characteristic" of that society'.

... Thus a society is more than a set of [actual] entities to which the same class-name applies: that is to say, it involves more than a merely mathematical conception of 'order'. To constitute a society, the class-name has got to apply to each member, by reason of genetic derivation from other members of that same society. The members of the society are alike because, by reason of their common character, they impose on other members of the society, the conditions which lead to that likeness.

It is evident from this description of the notion of a 'Society', as here employed, that a set of mutually contemporary occasions cannot form a complete society. For the genetic condition cannot be satisfied by such a set of contemporaries. Of course a set of contemporaries may belong to a society. But the society, as such, must involve antecedents and subsequents. In other words, a society must exhibit the peculiar quality of endurance. The real actual things that endure are all societies. They are not actual occasions. It is the mistake that has thwarted European metaphysics from the time of the Greeks, namely, to confuse societies with the completely real things which are the actual occasions. A society has an essential character, whereby it is the society that it is, and it has also accidental qualities which vary as circumstances alter. Thus a society, as a complete existence and as retaining the same metaphysical status, enjoys a history expressing its changing reactions to changing circumstances. But an actual occasion has no such history. It never changes. It only becomes and perishes. Its perishing is its assumption of a new metaphysical function in the creative advance of the universe. (AI, 203, 204)

The definition of "social order" in the above passage makes essential reference to "a common element of form" in the occasions of the society, and to "positive feelings" regarding that form. The "common element of form" does not refer to causal structure, but rather to a phenomenal feature that is ingredient in each occasion of the society. Likewise, the "positive feelings" involving that common element of form belong to the individual occasions. A society is not another entelechy with a subjective individuality of its own. Only individual occasions have feelings. The occasion is the only unit of experience. Whitehead's definition of "society" presupposes the phenomenological experience of each occasion in the causal order. Since an arrow diagram depicts only causal structure, it gives no indication that individual events have qualitative constituents. Thus, components essential to the definition of a society are not represented in the diagrams. The interpretation of arrow junctions as sentient occasions is required, as well as interpretation of the arrows themselves as prehensions of the phenomenal constituents of occasions by their immediate successors. A pattern of arrows may depict the causal structure of a society, but that society is only a society because every member inherits the defining phenomenal characteristics of that society from other members. The basic idea of a society is that it reinforces the satisfaction, or positive feelings, of its members. The *reason* for what is otherwise a meaningless pattern of arrows thus derives from the *reason* for the occurrence of individual events, which is individual satisfaction.

> The simplest example of a society in which the successive nexus of its progressive realization have a common extensive pattern is when each such nexus is purely temporal and continuous. The society, in each stage of realization, then consists of a set of contiguous occasions in serial order. A man, defined as an enduring percipient, is such a society. This definition of a man is exactly what Descartes means by a thinking substance. It will be remembered that in his *Principles of Philosophy* [Part I, Principle XXI; also Meditation III] Descartes states that

endurance is nothing else than successive re-creation by God. Thus the Cartesian conception of the human soul and that here put forward differ only in the function assigned to God. Both conceptions involve a succession of occasions, each with its measure of immediate completeness.

Societies of the general type, that their realized nexus are purely temporal and continuous, will be termed 'personal'. Any society of this type may be termed a 'person'. Thus, as defined above, a man is a person.

But a man is more than a serial succession of occasions of experience. Such a definition may satisfy philosophers— Descartes, for example. It is not the ordinary meaning of the term 'man'. There are animal bodies as well as animal minds; and in our experience such minds always occur incorporated. Now an animal body is a society involving a vast number of occasions, spatially and temporally coordinated. It follows that a 'man', in the full sense of ordinary usage, is not a 'person' as here defined. He has the unity of a wider society, in which the social coordination is a dominant factor in the behaviors of the various parts.

Also, when we survey the living world, animal and vegetable, there are bodies of all types. Each living body is a society, which is not personal. But most of the animals, including all the vertebrates, seem to have their social system dominated by a subordinate society which is 'personal'. This subordinate society is of the same type as 'man', according to the personal definition given above, though of course the mental poles in the occasions of the dominant personal society do not rise to the height of human mentality. Thus in one sense a dog is a 'person', and in another sense he is a non-personal society. But the lower forms of animal life, and all vegetation, seem to lack the dominance of any included personal society. A tree is a democracy. Thus living bodies are not to be identified with living bodies under personal dominance. There is no necessary connection between 'life' and 'personality'. A 'personal' society need not be 'living', in the general sense of the term; and a 'living' society need not be 'personal'. (AI, 205, 206)

The human sentient mind is one subsociety threaded into the causal structure of the living human organism. A typical conscious purpose arises from some bodily need, persists through a phase of pertinent goal-seeking bodily behavior, and subsides in a phase of fulfillment and satisfaction, with further benefits accruing to other bodily subsocieties in the process. The ultimate beneficiaries are always component individual occasions, since "feeling" and "well-being" apply only to such individuals. In the case of life processes of the body, the component occasions of each organ or suborganism profit from their associations with occasions of other organs in symbiotic fashion. The "associations" are traceable in principle to causal chains that lead from some member of one subsociety to some member of another. Conceivably, a single occasion could belong to more than one society, forming a causal conduit between societies. In any case, each causal transition is the unmediated influence of some completed occasion upon some other occasion in process of formation. Without some limiting factor, such as conflict of purpose between occasions, one would expect the world to rapidly arrive at an optimal structure in which harmony reigns and all occasions profit by the maximum synergy of their sharing arrangements. Whitehead accounts for the limitation of harmony by recognizing "negative prehensions," which are disagreeable relations between occasions. Prehensions are raw, intimate relations. In visceral terms, instances of negative prehensions might involve pain, disgust or repulsion. With prehensions as the basis of causal structure, we have a teleology of positive and negative feelings, which conforms to the traditional concept of "final causation." This contrasts with "efficient causation" as described by the purely structural formulations of physics. In the manner that structure presupposes component relations, efficient causation presupposes final causation, according to the hypothesis we are exploring. Efficient causation depends on the patterns of temporal sequence created by final causation.

It is one thing to reconcile efficient causation and final causation in a very general way, but this does not of itself yield any practical knowledge regarding cause-and-effect. Of specific concern to human beings is human psychology, where we cannot with any confidence sort out final causation from efficient causation. A psychological theory such as Freud's, which is framed in terms of purposeful agencies, employs a mode of explanation that is genuinely psychological. On the other hand, much of psychology is taken up with the chemistry and physics of the brain, which relies on a mode of explanation that is not psychological in the least. The two modes of explanation differ in the characterization of those occasions that do not belong to our conscious experience but are thought to constitute its immediate causes and effects.

In terms of efficient causation, the unknown causal agencies that interact with the conscious mind are generic quantum events, causally contiguous to our own sentient events in the cortical region of the brain. It is likely that those unknown brain events are close in temporal frequency to the human sentient events with which they interact. To physical science, no further conceptualizing is necessary. It is now entirely a matter of determining the temporal sequencing of the quantum-electrodynamics of the brain—that is, the causal order of quantum events, at electronic energy levels, in the cortical region of the brain. This describes psychology as "brain science" —a special case of physics, employing the scientific method.

In Freud's psychology, purposeful agencies are in conflict, resulting in the repression of some agencies to the unconscious. Here they remain active causal agents, using subversive tactics to achieve their purposes without arousing the conscious mind. Is this sort of theory in conflict with brain science?

With the assumption that individual events are inherently sentient and purposive, we should expect causal interaction with the human series at close quarters to be generally psychological in nature. It may well be that the dominant strand of human consciousness is one among many that vie for control of the body's resources, and that a great many of the unknown quantum events of the brain are organized into what Freud called "the unconscious." There is no inevitable clash between inherently psychological theories and the physicalistic theory of brain activity. A theory of the former type ventures an interpretation as to the intrinsic nature of events, while the physicalistic conception is free from such interpretation.

> The Universe achieves its values by reason of its coordination into societies of societies, and in societies of societies of societies. Thus an army is a society of regiments, and regiments are societies of men, and men are societies of cells, and of blood, and of bones, together with the dominant society of personal human experience, and cells are societies of small physical entities such as protons, and so on, and so on. (AI 206)

Science today thinks in terms of "big and "small." Galaxies are big and quanta are small. These are designations of spatial extent. They constitute a nasty distortion of the facts. Physical size and measure pertain to quanta. Quanta have greater or less *duration*, which means they are relatively slow or quick. The quicker quantum has greater energy. The slower quantum has a greater time span. In a vast closed region, a single quantum could connect the earliest moment to the latest, spanning an eon of time. Such a quantum has feeble energy, but to think of that quantum as "small" is to miss the fact that it spans a galaxy.

The electrons and photons, the rock, the brain, the planets, the galaxies-- all these have their quanta and their mass-energy. None of it has spatial extension because the chaining of temporal transitions is the only type of extension for physics— temporal extension. There is no such thing as a physical *state* of instantaneous organization. What exists all-at-once is only each individual moment. Since each moment (for physics) is generic and primitive, there is no specification of its state. To refer to a physical state, such as a "brain state," is to "freeze out time" and indulge in the illusion of instantaneous spatial extension.

We are apt to take a parochial view of our own native frequency as "just right" for the enjoyment of sane, coherent experience. We have trouble granting experience to the moments that comprise an electron or a proton because of their nanosecond quickness. But that pace is strictly relative to other frequencies. There is no absolute measure of duration. The pace of experience is "just right" for the constituent occasions of any sequence, regardless of its frequency ratio to other sequences.

> Nature is a complex of enduring objects, functioning as subordinate elements in a larger spatial-physical society. This larger society is for us the natural universe. There is however no reason to identify it with the boundless totality of actual things.

> Also each of these enduring objects, such as tables, animal bodies, and stars, is itself a subordinate universe including subordinate enduring objects. The only strictly personal society of which we have direct discriminative intuition is the society of our own personal experiences. We also have a direct, though vaguer, intuition of our derivation of experience from the antecedent functioning of our bodies, and a still vaguer intuition of our bodily derivation from external nature.

> Nature suggests for our observation gaps, and then as it were withdraws them upon challenge. ...

> Another gap is that between lifeless bodies and living bodies. Yet the living bodies can be pursued down to the edge of lifelessness. Also the functionings of inorganic matter remain intact amid the functionings of living matter. It seems that, in bodies that are obviously living, a coordination has been achieved that raises into prominence some functionings inherent in the ultimate occasions. For lifeless matter these functionings thwart each

other, and average out so as to produce a negligible total effect. In the case of living bodies the coordination intervenes, and the average effect of these intimate functionings has to be taken into account. ...

> Life may characterize a set of occasions diffused throughout a society, though not necessarily including all, or even a majority of, the occasions of that society. The common element of purpose which characterizes these various occasions must be reckoned as one element of the determining characteristics of the society. It is evident that according to this definition no single occasion can be called living. Life is the coordination of the mental spontaneities throughout the occasions of a society. (AI 206 207)

Life cannot be attributed to an individual occasion. Life only arises as a feature of societal organizations of occasions. Similarly, a single occasion is not intelligent. The momentary act of a single occasion is all intuition and feeling. Occasions do vary however, in the complexity of their internal phenomenological structure. The complexity of constituent occasions may be crucial to the intelligence level of their society, which is in turn needed to sustain a degree of complexity in the members. We humans can gauge the complexity of our own phenomenology directly. The complexity of our percepts prompts us to infer a corresponding complexity in the immediate external causes of those percepts. In general, when we progress in some particular domain of scientific investigation to the detailed level of individual events, we can imagine the qualitative richness of the constituent occasions to be commensurate with the number of causal relations impinging on each occasion according to the theory of its embedding causal structure.

To summarize this chapter, we start with a causal order of events as the basis of scientific theory. We then obtain a fitting interpretation for events and their ordering relations in "sentient occasions" and "prehensions." A "society" of occasions can then be defined, which supplies an explanation for certain persistent causal patterns. This interpretive scheme is then applied to the consideration of living and non-living processes of nature and the human organism in particular. Special care has been taken throughout to remain consistent with the causal analysis of the world into whole-and-part as depicted by the arrow diagrams, so that the interpretive scheme, and the purely structural import of scientific knowledge, reinforce and complete one another.

A primary objective for this chapter was to flesh out an interpretation of "causal relation." This comes down to an intuition of time and temporal passage. Whitehead can have the final word.

> [Plato] wrote in the Sophist, not-being is itself a form of being. He only applied this doctrine to his eternal forms. He should have applied the same doctrine to the things that perish. He would then have illustrated another aspect of the method of philosophic generalization. When a general idea has been obtained, it should not be arbitrarily limited to the topic of its origination.

> Thus we should balance Aristotle's—or, more rightly, Plato's—doctrine of becoming by a doctrine of perishing. When they perish, occasions pass from the immediacy of being into the not-being of immediacy. But that does not mean that they are nothing. They remain 'stubborn fact'....

> The common expressions of mankind fashion the past for us in three aspects-- Causation, Memory, and our active transformation of our immediate past experience into the basis of our present modification of it. Thus 'perishing' is the assumption of a role in a transcendent future. The not-being of occasions is their 'objective immortality'. A pure physical prehension is how an occasion in its immediacy of being absorbs another occasion which has passed into the objective immortality of its not-being. It is how the past lives in the present. It is causation. It is memory. It is perception of derivation. It is emotional

conformation to a given situation, an emotional continuity of past with present. It is a basic element from which springs the self-creation of each temporal occasion. Thus perishing is the initiation of becoming. How the past perishes is how the future becomes. (AI, 237, 238)

Acknowledgements

It may have been Buckminster Fuller's little book *I Seem To Be a Verb* that started the wheels turning for me. "I seem to be a verb, an integral process of the universe." I got a spooky feeling from that book, which may have been the demise of belief in my self as a mental substance. That may be a prerequisite to making any progress on the mind-body problem. Then it was *Science and the Modern World* that struck me full force with the problem itself. A course given by Gerald Weiss at Macalester College, "History and Systems of Psychology," was most stimulating, and it was Mr. Weiss who started me reading Freud, Russell, and Gustav Bergmann. I must thank my college friend John Simpson for puzzling over the problem with me for the remainder of that last year at Macalester College. Also, mathematics professor Joseph Konhauser would post a "Problem of the Week" on the department bulletin board, and I often walked away with the 50-cent prize for best solution, which bolstered my confidence in my problem-solving ability.

After graduation, I immersed myself in general reading for a year, but remained thoroughly vexed by the problem. John Simpson had gone to graduate studies at the University of Iowa, and it turned out that his office was right next to that of Gustav Bergmann, whom by now I revered as a god. John said he could introduce me, so I drove down there. Mr. Bergmann asked me what I was reading, and I told him Russell's *Principles of Mathematics*. He remarked that it was a good book. I said I was interested in studying under him, but he told me that the University of Minnesota had a fine department, so why not enroll there?

I applied to graduate school at the U of MN. It happened, fortuitously, that I was too late to be considered for the coming year, and I would have to wait a year. I had been a math major with no formal background in philosophy, and it was unlikely I would have been accepted. I moved near the campus, got a cab-driving job, and looked for an extension course to fill the time. I selected a course called "Philosophy of Science," taught by Grover Maxwell. The text he had assigned was Russell's *Human Knowledge*. Reading that book did the trick. I think Grover pulled some strings, and I was accepted into the graduate program the next year.

In graduate school, I made friends with Mark Ward, another incoming student who had studied neurophysiology as an undergraduate at Stanford. Grover was our advisor. The three of us shared an enthusiasm for Russell as solver of the mind-body problem. Oddly enough, we struggled in vain to win anyone else over to this view. Grover was highly respected in the academic world as a philosopher of science, and he was co-editor of the leading periodical in that field. Herbert Feigl was also on the faculty at that time. I enjoyed my "brush with greatness" in making the acquaintance of these professors. I eventually concluded though, that I was not going to succeed where Grover was failing, and I was not ready to write a thesis. Content with having found the solution for myself, I dropped out and went to electronics school. I figured the solution would inevitably work its way into the light of day without my help.

It didn't though. I quit reading philosophy but kept tabs on things by scanning the magazine racks over the years. *Scientific American* came out with an end-of-millennium edition devoted to "Consciousness and the Brain." I concluded from this issue that the solution had not surfaced in our popular culture, and I resolved to write it up in no uncertain terms. I had come to appreciate the elusiveness of achieving technical clarity through thirty years of writing maintenance manuals and user documentation. I spent a year writing the preface and synopsis. I used the gift of unemployment to finish the rest. I'm personally quite satisfied, and I dare say the thing can't be made any clearer. Grover, I wish you hadn't died, because I've finally got the thesis I owe you.

www.ingramcontent.com/pod-product-compliance
Lightning Source LLC
LaVergne TN
LVHW040149080526
838202LV00042B/3085